T0260076

GAMES *for* YOUR MIND

GAMES
for
YOUR MIND

*The History and Future
of Logic Puzzles*

JASON
ROSENHOUSE

PRINCETON UNIVERSITY PRESS
PRINCETON & OXFORD

Copyright © 2020 by Princeton University Press

Requests for permission to reproduce material from this work
should be sent to permissions@press.princeton.edu

Published by Princeton University Press
41 William Street, Princeton, New Jersey 08540
6 Oxford Street, Woodstock, Oxfordshire OX20 1TR

press.princeton.edu

All Rights Reserved
ISBN 978-0-691-17407-5
ISBN (e-book) 978-0-691-20034-7

British Library Cataloging-in-Publication Data is available

Editorial: Susannah Shoemaker and Kristen Hop
Production Editorial: Brigitte Pelner
Jacket Design: Michel Vrana
Production: Jacqueline Poirier
Publicity: Matthew Taylor (US) and Katie Lewis (UK)
Copyeditor: Cyd Westmoreland

Jacket Images: iStock

This book has been composed in Sabon

Printed on acid-free paper ∞

Printed in the United States of America

1 3 5 7 9 10 8 6 4 2

To Noah and Sonia

The escape room was fun.
Consider this a continuation.

CONTENTS

PREFACE

Novelists often describe the experience of having their stories go in directions entirely different from what they had in mind when they sat down to write. They might create the characters and contrive a scenario for them, but from there, the characters are just going to do whatever it is in their nature to do, regardless of any preconceived notions the writer brought to the project.

To my surprise, I had a similar experience while writing this book.

My original intention had been for a relatively short, light-hearted book about logic puzzles. There are two towering figures in the history of recreational logic—Lewis Carroll (better remembered as the author of *Alice in Wonderland* and *Through the Looking-Glass*) and Raymond Smullyan. I figured I would present a selection of their puzzles with some historical and mathematical context, and then close with some puzzles based on nonclassical logics that I had devised myself.

What I had not anticipated was just how difficult it is to draw a clear line with amusing puzzles on one side, and difficult mathematical and philosophical questions on the other. Lewis Carroll explicitly integrated his puzzles into more serious, scholarly work on logic. He also published two academic papers in logic, but wrote them in the style of short stories with humorous dialog. Is that serious scholarship or merely recreational math? Raymond Smullyan saw his puzzles about knights (who always tell the truth) and knaves (who always lie) as a pedagogical tool for introducing readers to deep questions of mathematical logic, especially those surrounding Gödel's two famous theorems. Meanwhile, the very idea of "nonclassical" logic will sound strange to many people, since it is generally thought that logic is logic and that is all there is to it. For people not immersed in this subject, the idea that "logic" might require an adjective to clarify the sort of logic that is intended will seem strange. (Interestingly, the spell-checker on my computer insists that "logics" is not a word.)

Moreover, the more I delved into the literature, the more I noticed that the puzzles I was discussing were a microcosm of the history of logic generally. Carroll's puzzles focused on the ancient system of logic pioneered by Aristotle, which dominated the subject for most of its history. Smullyan's elementary puzzles explored propositional logic, which can be seen as a generalization of the Aristotelian system, while his more advanced puzzles explored the mathematical logic that supplanted Aristotle at roughly the turn from the nineteenth century to the twentieth century. Nonclassical logic has existed as a serious object of study since roughly the 1920s, but it really took off with the advent of computer science. It is a hot topic of study in the modern field of automated reasoning, which involves developing computers that can manipulate large, and sometimes contradictory, data sets.

The result was a much bigger, and hopefully much richer, book than I originally had in mind. Regardless, I have tried to write at a level that will be accessible to a general audience, while not doing too much damage to the sometimes difficult questions I found it necessary to discuss. It is almost inevitable that I have made some errors, and I am sure there are places where some philosophers will not agree with my conclusions. Hopefully, though, I have at least managed to provide some food for thought.

(Incidentally, my admission that there are likely to be some errors might seem a strange one. On one hand, I obviously think I have good reasons for believing every claim I make in the book, while on the other hand, I am acknowledging that the conjunction of all those claims is likely to be false. But if you rationally believe statements p and q individually, does that not imply you should also believe the statement "p and q"? This is known as the Paradox of the Preface, and we discuss it in Section 16.3. It is one of many examples of logic puzzles arising in the most unexpected places.)

The book is structured as follows. In the first part, I provide a general introduction to both logic and logic puzzles. The second part focuses on the work of Lewis Carroll. Chapter 3 provides a primer on Aristotelian logic. Chapter 4 considers Carroll's short book *The Game of Logic*, in which Aristotelian logic is presented as a game suitable for children. Chapter 5 then considers Carroll's longer book *Symbolic Logic*, which truly straddles the line between enjoyable puzzles and serious scholarship. I close this part by discussing Carroll's two academic papers in logic in Chapter 6.

In Part III, the focus changes to Raymond Smullyan. Chapter 7 introduces propositional logic and provides a sampling of puzzles about liars and truthtellers. We then move on to three chapters about mathematical logic. Chapter 8 provides historical context, and Chapter 9 explains a few important concepts. These chapters are background for Chapter 10, in which we discuss how Smullyan used liars and truthtellers as a device for illustrating Gödel's theorems. Chapter 11 closes out this part of the book with a discussion of puzzles centered around asking clever questions.

To this point, we have been discussing the history of logic puzzles. The possibilities for puzzles based on classical logic have been thoroughly explored by such writers as Carroll and Smullyan. Nowadays, however, it is routine for scholars to investigate systems of nonclassical logic. Puzzle creators need to keep up! The future of logic puzzles is found, I suggest, in crafting puzzles based on nonclassical logics. In Part IV we consider a few possibilities.

Chapter 12 introduces this subject, and Chapter 13 considers what life would be like if Smullyan's knights and knaves employed a multivalued logic. The puzzles in this chapter are my own creations.

Finally, Part V rounds up a few miscellaneous topics that did not fit well in the other chapters. Chapter 14 discusses the so-called "Hardest Logic Puzzle Ever," which was introduced by philosopher George Boolos in 1996. Since then, it has produced a small industry of papers discussing its various nuances and cheekily offering up ever-more difficult versions of the puzzle. Chapter 15 discusses a genre known as "metapuzzles." These are puzzles that can be solved only by knowing whether certain other puzzles could be solved. Chapter 16 considers a selection of paradoxes, which again straddle the line between amusing puzzles and difficult philosophical questions. Chapter 17, the final chapter, goes off in a different, and lighter, direction from what has preceded it. I introduce the term "logic fiction" to refer to a category of literature in which the main interest lies in the impressive feats of logical deduction undertaken by the protagonist. Specifically, I am referring to the so-called classical detective story. These works are logic puzzles in the form of novels, and they fully deserve to be included here. I provide a brief history of the genre and also a small selection of a few of my favorite works. I hope that you will enjoy reading this as much as I enjoyed writing it.

While most of the chapters will be accessible to anyone willing to make the effort, a few will prove more challenging. Specifically, Chapters 3, 10, and 14 discuss fairly technical subject matter. Even here, however, I have tried to write as clearly as possible and to at least make the main ideas come through even when the details get complex or tedious.

I express my deepest gratitude to two anonymous reviewers, who made various helpful comments on earlier drafts of this book. I also thank Vickie Kearn, Susannah Shoemaker, and everyone else at Princeton University Press for their extraordinary patience and encouragement during the writing of this book. The manuscript was well over deadline by the time I handed it in, but they never pressured me to speed things up.

Let me close on a personal note. My father used to challenge me with logic puzzles when I was a kid, and I have loved them ever since. I especially remember him showing me the problem presented here as Puzzle 89 (in Chapter 15), at an age where I was not yet able to make heads or tails of it. He steadfastly refused to tell me the solution, or even to give me a hint, insisting that I solve it for myself. This I eventually did, albeit years later. After completing graduate school and starting a career as a mathematician, I had the opportunity to spend a weekend with Raymond Smullyan at his home in upstate New York. I could not have asked for a more gracious host. I hope that this book will pay forward much of the pleasure and satisfaction these puzzles have, over the years, given me.

PART I
The Pain and Pleasure of Logic

CHAPTER 1

Is Logic Boring and Pointless?

1.1 Logic in Practice, Logic in Theory

Logic is easy in specific cases, but difficult in general.

If I tell you that all cats are mammals and that all mammals are animals, then you conclude that all cats are animals. If another time I tell you that my cat is always asleep at 4 o'clock in the afternoon, and you then notice that it *is* 4 o'clock, then you conclude that my cat is asleep. If instead you see my cat walking around and plainly not asleep, then you conclude that it is not 4 o'clock.

That is logic.

This is simple. These conclusions are *obvious*. Suppose, though, that someone doubts you. He asks, "How do you *know* it follows that all cats are animals? How can you be *certain* that the cat is sleeping or that the time is 4 o'clock?" You would hardly know what to say. The fundamental principles of logic seem so straightforward and intuitive that it is unclear how to explain them in terms of something more readily comprehended. You would assume that the skeptic in some way misunderstands the language. You might even repeat the premises to him, slower and louder.

The ancient philosopher Sextus Empiricus, writing in the second century CE and commenting on the work of the Stoic logician Chrysippus (279–206 BCE), suggested that even dogs understand these principles:

[Chrysippus] declares that the dog makes use of the fifth complex indemonstrable syllogism when, on arriving at a spot where three ways meet. . ., after smelling at the two roads by which the quarry did not pass, he rushes off at once by the third without stopping to smell. For, says the old writer, the dog implicitly reasons thus: "The animal went either by this road, or by that, or by the other: but it did not go by this or that, therefore he went the other way." (Floridi 1997, 35)

There really does seem to be something instinctive about the principles of logic. Every time you search for your lost keys by retracing your steps, you are applying those principles. You say to yourself, "I know I had the keys when I left the house. I then visited locations A, B, and C, and I could only have left the keys at a location that I visited; therefore, I left the keys at one of locations A, B, or C." Of course, you never really pause to spell out the steps of the argument, and that is precisely the point. You process the basic logic of the situation so automatically you are hardly aware that you have reasoned at all.

The extreme naturalness of logical reasoning was noted by John Venn, in his 1881 book *Symbolic Logic*:

> It may almost be doubted whether any human being, provided he had received a good general education, was ever seriously baffled in any problem, either of conduct or of thought...by what could strictly be called merely a logical difficulty. It is not implied in saying this, that there are not myriads of fallacies abroad which the rules of Logic can detect and disperse.... The question is rather this: Do we ever fail to get at a conclusion, when we have the data perfectly clearly before us, not from prejudice or oversight but from sheer inability to see our way through a train of merely logical reasoning? ... The collection of our data may be tedious, but the steps of inference from them are mostly very simple. (Venn 1881, xix)

As Venn notes, people do commit logical errors. In my role as a mathematics educator, I encounter such errors all the time. For example, it is common for students to treat a statement of the form, "If p, then q," as though it were logically equivalent to the statement, "If q, then p." Rarely, though, is the student confused about abstract principles. If you point out to him that the statement "If Spot is a normal dog, then Spot has four legs" is true, while the statement "If Spot has four legs, then Spot is a normal dog" is false, he immediately understands your point. He does not argue that *he* is the one who is thinking clearly, while *you* are the one who is confused. Such errors occur simply because mathematicians deal with complex statements about abstract objects, and, in the heat of the moment, students often find them difficult to parse.

Returning to my opening examples, the conclusions not only followed from the premises, they followed in a particular way. Consider a contrasting example. Recently I was preparing dinner for a large group of friends. I was in the kitchen, with the exhaust fan, the sink, and the television all going. It was *noisy*. Suddenly, one of my cats came barreling in, her paws struggling for purchase on the smooth kitchen floor. She darted down the basement stairs. I reasoned, "My cat only panics like that when she hears strangers on the patio. I'll bet my guests have arrived, but I did not notice the doorbell amidst all the racket." I went to the door and found that I was right.

I came to this conclusion because I knew many instances of my cat panicking at the sound of strangers on the patio, and no instances of her panicking in that manner in response to any other stimulus. I reasoned from empirical facts and extensive personal experience to the conclusion that my guests had arrived. Philosophers would say that my reasoning was *inductive*, from Latin words that translate roughly as "to lead into." In this case, I was led to a general conclusion from my experiences in a few specific cases. This style of reasoning is common in science, where our confidence in a theory's correctness grows each time it accurately predicts the outcome of an experiment.

Our opening examples were not of that sort. From "All cats are mammals" and "All mammals are animals," we concluded that "All cats are animals," but this conclusion in no way followed from anything we know about cats, mammals, or animals. From "All cats are green" and "All green things are plants," it follows that "All cats are plants," though in this case, all three statements are false. Our notions of what follows logically from what are unrelated to the facts of the world. Instead, they are related in some manner to the way we use language and to the grammatical structure of the assertions involved. We could say "All As are Bs, all Bs are Cs, and therefore, all As are Cs" without causing controversy.

This sort of reasoning is known as *deductive*, from Latin words meaning "to lead down from." It is the primary sort of reasoning employed in mathematics. Deductive reasoning seems to have a certainty about it that inductive reasoning lacks. My conclusion that my guests had arrived given the evidence of my panicked cat was perfectly reasonable. However, it might have been that my cat had been scared by something else, or was just being weird for some reason. My conclusion *might* have been wrong. But if the statements "All cats are mammals" and "All mammals are animals" are both true, then it simply *must* be true that "All cats are animals." Period. End of discussion.

This all seems sufficiently straightforward. The difficulty comes when trying to formalize our intuitive notions. Can we write down a general set of rules to tell us what follows from what?

We have seen that logical inferences are closely related to language, and, indeed, "logic" comes to us from the Greek "logos," meaning "word." In natural languages—English, French, German, and so forth—there are many types of words. There are nouns, which we can take to represent objects, and verbs, which generally describe what the nouns are doing to each other. There are adjectives to supply additional information about the nouns, prepositions to describe relationships among them, and adverbs to tell us more about the verbs.

Then there are other words whose function is to establish logical relationships among the component clauses of a complex sentence. Philosophers refer to these words as *logical constants*. In English, we use such words and constructions as "not," "and," "or," and "if–then" as logical constants. You come to understand the meanings of these words by understanding the effect they have on the clauses to which they are connected.

For example, if I say, "On Tuesday, I ate cookies, and I ate cake," the role of "and" is to tell you that I ate both cookies and cake on Tuesday. If you later discover I only ate one of them, or neither of them, you would think I had said something false. In this context, we come to understand what "and" means by understanding the truth conditions it imposes on the sentence whose clauses it is joining. Moreover, "and" plays this role regardless of the content of the clauses on either side of it. That is why it is called a "logical constant."

This is progress toward our goal of having general rules for telling us what follows from what. If we let p and q represent simple assertions, then we can say that from the sentence "p and q" we can fairly conclude that p and q are both true individually. Given some familiarity with standard English usage, we can quickly write down other such rules:

- The statement "not p" has the opposite truth value from p.
- Given "p or q," and "not p," we can conclude that q is true.
- Given "If p, then q," and "p," we can conclude that q is true.

There is much that could be added to this list, of course. For the moment, however, the main point is that this logic business does not seem very complicated at all. Writing down logical rules involves nothing more than understanding what words mean, and you hardly need a degree in philosophy for *that*.

Matters are not always so simple, however. If I am at a restaurant, the server might ask me whether I want french fries or mashed potatoes with my dinner. Later he might ask me whether I want coffee or dessert. In the first instance, it is understood that I am to choose only one of french fries or mashed potatoes, while in the second it would be acceptable to have both coffee *and* dessert. What, then, should the rule be for statements of the form "p or q"? If p and q are both true individually, should "p or q" be deemed to be true? Or is it false? It would seem there is no rule that covers all contexts.

And how are we to handle conditional statements, by which I mean statements of the form "If p, then q"? If p is true by itself, and q is false by itself, then "If p, then q" should be considered false. *That* much is clear.

But what if p and q are both true? Should we automatically declare "If p, then q" to be true in this case? That seems reasonable for mathematical statements: "If x and y are even numbers, then $x + y$ is even as well," for example. In contrast, what am I to make of the statement, "If I am not a cat, then I am not a dog"? Both parts are true by themselves, but the sentence as a whole does not seem to be true. In everyday usage, we normally take it for granted that the two parts of a conditional statement are relevant to each other, but it is unclear how to capture a relevance requirement in a logical system.

Natural languages have many other attributes that make logical analysis very difficult. They contain statements that are vague or ambiguous. Some statements are *indexical*, which is to say that their meaning depends on the context. For example, the meaning of "I am hungry" changes, depending on the speaker. The truth or falsity of a statement often depends on more than just its grammatical structure. For example, the statements "If my cat did not eat the tuna, then someone else did" and "If my cat had not eaten the tuna, then someone else would have" have very different meanings, though we might naively interpret both as having the abstract form "If p, then q."

It would seem that trying to capture the logical rules implicit in everyday language is not so simple after all.

Seeking respite from such travails, logicians prefer instead to work with formal languages. By a "formal language," I mean a language the logician simply invents for her own purposes. The logician therefore has complete control over what counts as a proper assertion, and she can devise strict rules for determining the correctness of proposed inferences. There is no vagueness and no ambiguity. For logicians, the move from a natural to a formal language produces a calming effect, similar to when the kids are out for a few hours and blessed quiet descends on the house.

In crafting her language, the logician might begin by inventing symbols to represent basic sentences. Other symbols are then devised to denote familiar connectives, like "and," "or," and "if–then;" and still more symbols are introduced to denote various sorts of entailments and implications. As a result, simple assertions can be made to look complex. For example, our inference that all cats are animals from the assumptions that all cats are mammals and all mammals are animals, might end up like this:

$$(\forall x \ Cx \to Mx) \land (\forall x \ Mx \to Ax) \models (\forall x \ Cx \to Ax).$$

You should interpret "Cx," "Mx," and "Ax" to mean, respectively, that x is a cat, mammal, or animal. The upside down A is an abbreviation of "for all," the arrow means "if–then," and the vertical wedge means "and."

The symbol that looks like the Greek letter pi on its side denotes entailment. Thus, translated back into English we have, "The assumptions that for all x, if x is a cat, then x is a mammal, and for all x, if x is a mammal, then x is an animal, entail the conclusion that for all x, if x is a cat, then x is an animal."

Practitioners of formal logic are fond of this sort of thing. A statement as simple as "My cat is furry" might be rendered thus:

$$(\exists x) \, (Jx \wedge (\forall y) \, (Jy \rightarrow (y = x) \wedge Fx)) \, .$$

In English, this collection of symbols means: "There exists an x such that x is Jason's cat, and if y is anything else that is Jason's cat, then y is the same as x, and y is furry." Where you might see a simple statement of fact about my cat, a logician sees a complex existential assertion involving conditional statements and conjunctions. This, from a sentence containing neither the word "and" nor "if–then." It would seem that a difficult logical structure of language lurks beneath its grammatical structure.

The relationship of the formal language to natural language is like that of a laboratory experiment to the real world. Scientists contrive controlled scenarios in which a few variables can be studied in isolation from others. They then hope that they have chosen the really important variables, so their results will be applicable to reality. Likewise, the logician hopes that the formal language captures those aspects of natural language that are relevant to reasoning, even though she knows subtle aspects of the natural language will inevitably be lost in her formalization.

1.2   *Enter the Philosophers*

The translation of simple, natural-language sentences into difficult symbolic ones can be a tedious affair, but the worst is still to come. Once the philosophers learn of your project, they will want a piece of the action, and God help you when *that* happens. Philosophers have investigated, *minutely*, all of the central notions on which logic relies. Through their investigations, they have discovered the only thing philosophers ever discover: that everyday notions used without incident in normal social interactions become murky when closely analyzed.

For example, most elementary textbooks will tell you that the fundamental unit in logic, comparable to atoms in physics or prime numbers in arithmetic, is "the proposition." If we ask, "What sort of thing is it that can rightly be described as either true or false?" the answer is, "A proposition." It is gibberish to say, "This vegetable is true" or "This color is false," but it makes perfect sense to say, "This proposition is either true or false."

But what *are* propositions?

One possibility is that a proposition is just the same thing as a declarative sentence. This seems plausible. In a conversation, we might say, "What you just said? That's so true!" when what the person just said is actually a sentence. High school students often take examinations in which they are asked to mark each of a list of sentences as being either true or false. So, maybe the concept of "proposition" does not really add very much, and we should just talk directly about sentences instead.

The problem is that the same sentence can mean different things in different contexts. When I say, "I have a cat," I am not expressing the same proposition as you are when you utter those same words. However, two different sentences might express the same proposition. In France, I would say, "J'ai une chat" instead of "I have a cat," but the same proposition has been expressed. Some sentences do not seem to express any proposition at all. "It is raining" is a perfectly fine sentence, but until we contextualize it to a time and a place, we cannot assign it a truth value. In light of these considerations, it seems accurate to say that we use declarative sentences to express propositions, but that the sentences are not themselves propositions. There are concepts of some sort to which sentences point, and *those* are the propositions.

Moreover, propositions do not just get stacked up into written arguments so that other propositions might be drawn as conclusions from them. They have another existence as beliefs in a person's mind. When you believe something about the world, what kind of thing is it that you actually believe? A proposition, that's what! However, it does not seem right to say that the thing you believe is a sentence, as though you cannot have beliefs unless you have first summoned forth sentences that express them. My cat has beliefs about the world, but, clever though she is, I doubt that she can express those beliefs in sentences.

So the question persists. What *are* these things we call "propositions"? Are they just the meanings of sentences? Can we define "proposition" as "what a sentence means"? Perhaps, but does this really help us understand what is going on? Meaning is itself a very difficult concept, as pointed out by philosopher A. C. Grayling in a discussion of this very point:

> Suppose I am teaching a foreign friend English, a language of which he is wholly ignorant; and suppose I point to a table and utter the word 'table.' What settles it for him that I intend him to understand the object taken as a whole? Why should he not take me as pointing out to him the colour, or the texture, or the stuff of which the object is made? Imagine my pointing at the table-top and saying 'glossy.' Why should he not understand me as naming

the object as a whole, rather than the style of its finish? At what is apparently the simplest level of demonstratively linking a name with the object it is supposed to 'mean,' then, there are puzzling difficulties. (Grayling 1982, 36)

If meaning is difficult even in this simple case, then how much more difficult is it when we speak of the meaning of a whole sentence? For example, how is understanding the meaning of a sentence different from just understanding the proposition it expresses?

It is at this point, when most people find their eyes glazing, that the philosophers start to get *really* interested. Their chief weapon in the fight against vagueness and imprecision is the drawing of subtle distinctions, and the literature in this area offers plenty of them: between sentences, statements, and propositions; between sense, meaning, and reference; between the intension and the extension of a term. Not to mention what is potentially the most important distinction of all: between realism and nominalism with respect to abstract objects. You see, if you take the view that there are these spooky, ill-defined propositions floating around just waiting to be gestured at by sentences, then you sure seem to be suggesting that abstract objects actually exist. That makes you a realist. Against you are the nominalists, who regard abstract objects as useful fictions that humans devise for their own purposes. (Does the number three actually exist as an object by itself? Or is "three" just a name we use to describe what is common among all collections of three physical objects?) This particular dispute has raged for centuries, and I assure you that the rival camps see this question as *very* important.

Do you see what happened? We asked, in perfect innocence, what propositions were, and just a few paragraphs later, we were mired in deep questions of ontology and metaphysics. For heaven's sake.

Let us put these niceties aside. Assume for the moment that we have arrived at a coherent account of "proposition." What does it mean for this proposition to be "true"?

Any nonphilosopher would say that the true propositions are the ones in accord with the facts. We have facts on one side, true propositions on the other, and for every true proposition, there is a corresponding fact that makes it true. What could be simpler?

We could retort, however, that this approach is *too* simple. A philosopher might say, "Yes, thank you, I *know* that truth is about correspondence with facts in some vague way, but that is unhelpful. I need to understand the process by which a proposition is paired with the fact to which it corresponds. If I asked, 'What caused this patient's death?' you would no doubt reply, 'He died because his heart stopped,' thinking you had thereby said something informative. But the question, *obviously*, is what caused his heart to stop. Likewise, the question for those claiming

Figure 1.1. The empirical fact corresponding to the proposition, "My cat is watching me type this."

that truth is about correspondence with fact is to explain the nature of this correspondence, and good luck with that."

How *do* we go unerringly from the true proposition to its corresponding fact? Correspondence seems straightforward when considering simple assertions. "My cat is watching me type this" is true because of a certain empirical fact, depicted in Figure 1.1. Matters are far less straightforward when discussing complex statements. What is the fact of the world corresponding to "If my cat had not broken her leg, I would have spent Saturday either reading a book or watching television, instead of rushing her to the veterinarian"? It would seem that facts can be rather complex. To make the correspondence theory work, I would first need an account of what facts are and then an account of the manner in which the pairing of true propositions with facts is achieved. Neither of these accounts is readily forthcoming.

Other types of sentences cause problems as well. What fact of the world corresponds to "There are no unicorns"? Perhaps the relevant fact is found by restating the sentence in the equivalent form: "Everything is a non-unicorn," but, among other problems, this suggests that a sentence that certainly appears to be about unicorns is actually about literally everything except unicorns. Similar problems could be adduced for disjunctions (or-statements), counterfactuals, statements about the past, and statements about abstract entities (like $2 + 2 = 4$). In each case, it is not straightforward to identify the piece of reality to which the proposition corresponds.

The more you think about it, the more difficult it becomes to pin down the correspondence relation that is said to obtain between true propositions and facts. Propositions are abstract entities, some notion of which resides in our heads. Facts are about physical objects that exist out there in the world. "Correspondence" implies some sort of isomorphism between these radically different realms. How can that be?

Perhaps you think the solution is as follows. We begin by identifying certain simple, basic facts. These correspond straightforwardly with simple propositions, by which we mean propositions with no logical structure to them. The orange cat on my sofa corresponds simply to the proposition, "My cat is orange." The facts corresponding to more complex statements are then found by breaking the statements down into the logical simples out of which they are made. Done!

The philosophers have a name for this approach, which is never a good sign. It is called "logical atomism," the idea being that these logical simples are like the atoms out of which chemical substances are made. At various times, this approach has been defended by giants like Bertrand Russell, Ludwig Wittgenstein, and Rudolf Carnap. Nowadays, however, the notion has fallen on hard times, for reasons you have probably already guessed. Those logical simples have proved surprisingly elusive, and no one has managed to supply a helpful account of them.

It would seem that the correspondence relation is so murky and complex that we might reasonably wonder whether it is actually helpful in elucidating the nature of truth. The main argument in favor of the correspondence theory (really, the only argument) is its agreement with common sense. In daily life, it sure feels like we assess truth first by understanding a sentence's meaning and then by comparing it with relevant facts. Philosophers, though, take special delight in refuting common sense. Tell a philosopher that an idea is intuitively obvious, and he will quickly retort that, so sorry, it is incoherent nevertheless.

At this point, we might think that our whole model is wrong. We have been acting as though we have the world of propositions *over here*, and then separately from that, there is an objective reality *over there*. This objective reality comes equipped with facts, and in some vague way, it is these facts that make propositions true. The relation between fact and proposition is said to be one of correspondence, but we encountered difficulty spelling out the nature of this relation.

There are other possibilities. Maybe it is not facts (whatever *they* are) that make propositions true, but rather other propositions. That is, we could say that a proposition is true when it coheres with other propositions that are already accepted. Defenders of this view argue that the relation of coherence is more readily described than that of correspondence. Or maybe the whole concept of truth is just redundant.

After all, what is the difference between saying, "Proposition p is true" and just asserting p in the first place? In this view, stating that a proposition is true is different from stating that an apple is red. The latter case attributes a property to an object, while the former does not. These are called the "coherence" and "redundancy" theories of truth, respectively. They have their defenders, as do several other theories I have chosen, because I want people to keep reading my book, to omit.

Mighty treatises and mountains of journal articles have been written on each of these matters, and believe me when I tell you, they do not make for light reading. Nothing to relax with before bed in *that* charming little ocean of verbiage. Perhaps, though, we are justified in ignoring this literature. Just as I can drive a car without knowing how it works, so, too, can I use notions like "proposition" and "truth" without a proper philosophical account.

Sadly, though, we are just getting started. Once you start asking philosophical questions about logic, it is impossible to stop. Do the laws of logic exist by necessity, are they just arbitrary consequences of the way we define words, or are they empirical facts discovered through investigation and experiment? Should logicians be seeking the one true logic that applies always and everywhere? Or are systems of logic more like systems of geometry: useful or not useful in different contexts, but not correct or incorrect in any absolute sense? Should "true" and "false" be regarded as the only truth values? Some statements are vague, after all, and therefore do not fit comfortably into a binary conception of truth. To accommodate this fact, perhaps we should countenance truth-value gaps, by which I mean propositions that are neither true nor false. Perhaps we want a third truth value, "neutral," which applies to statements that are vague. Maybe we should countenance the possibility of truth-value gluts, as when we find that a proposition is true in one sense but false in another. Maybe "both true and false" ought to be an option. How should we handle different modalities? Some propositions are possibly true, as when I say, "Tomorrow I will go to the park," while some are necessarily true, like "Two plus two equals four." Should our system of logic reflect this difference?

No point about logic is too clear and simple to avoid the complexifications of philosophers. Delve into this literature, and you will encounter careful discussions of the distinction between the philosophy of logic and philosophical logic. You will see people get the vapors over the problem of assigning a truth value to the sentence: "This very sentence is false." (If it is false, then what it asserts is actually the case, so it is true. But if it is true, then what it asserts must be the case, which makes it false!) You will meet people of unquestioned brilliance and sagacity thinking that it is an insightful commentary on the nature of truth to observe that "Snow is white" is true if and only if snow is white.

And you will encounter people who argue, in perfect seriousness, that some contradictions are true. You read that right. No idea is so daft that some philosopher has not floated it.

Folks, *this* is what awaits you if you choose to study logic. Textbooks on formal or mathematical logic appear to be written in hieroglyphics. They are also often disappointing, in that after translating the symbols into English, it is common to find that something trivial has been asserted. Texts on philosophical logic are well-nigh unreadable, and often leave you with the uncomfortable feeling you know less after reading them than you did before. Whatever the precise subject matter, if the word "logic" is in the title, it is likely not the sort of book most people would enjoy reading. Taking a course in the subject would rank low on almost anyone's bucket list. "Recreational logic," in this view, must simply be dismissed as an oxymoron.

Which is a pity, since deducing the logical consequences of a set of premises can be surprisingly enjoyable.

1.3 Notes and Further Reading

The philosophy of logic is an inherently difficult subject, and even its introductory texts make for heavy reading. I found the books by Haack (1978), Grayling (1982), and Read (1995) to be especially helpful. Sarcasm aside, I found all three books to be fascinating, even if I was occasionally unpersuaded of the importance of some of the more esoteric discussions. While it is true than one can drive a car without understanding how a car works, I am certainly happy there are people out there who *do* understand how cars work.

The correspondence theory of truth has recently been the subject of a book-length defense by Joshua Rasmussen (2014). In particular, he provides a detailed discussion of the sentence, "There are no unicorns." Both this book and the references contained therein will be helpful to anyone wanting to look into this area.

The view that some contradictions are true is referred to as *dialetheism*. It is definitely a fringe view among logicians generally, but it is taken seriously and has enthusiastic defenders. At one point, I considered trying to devise some dialetheistic logic puzzles for this book, but ultimately I found the idea just a little too unnatural to wrap my head around. At any rate, philosopher Graham Priest has been the most eloquent champion of the idea, and you can check out his books *In Contradiction* (2006) and *Doubt Truth to Be a Liar* (2008a) for more information.

CHAPTER 2

Logic Just for Fun

2.1 Sudoku and Mastermind

If you ask the average person what he thinks of mathematics, there is a good chance his reply will not be printable. People not only claim to dislike mathematics, they typically do so with pride and gusto. Often, though, the issue is really one of presentation. If you ask someone to find all solutions in positive whole numbers to the system of equations

$$x + y = 50,$$
$$2x + 3y = 120,$$

you are in danger of dredging up unpleasant memories from high school algebra class. Instead, suppose you tell him that you have 120 treats to divide among a collection of 50 kittens and puppies, and that each puppy gets three treats while each kitten gets two (because the puppies are bigger and hungrier). Then ask them how many puppies and how many kittens there are. The response is likely to be a bit more favorable. You could also note that no algebra is necessary to solve this: Just imagine giving every animal two treats. That takes care of 100 treats, leaving 20 left over. Each puppy has one more treat coming to him, which means there must be 20 puppies. That leaves 30 kittens. At this point, that person who moments ago boasted of his eternal hatred of mathematics might just be smiling.

Something similar happens with logic. The same person who claims she would only handle a logic textbook while wearing gloves is often one who enjoys a good logic puzzle. How else to explain the popularity of Sudoku?

3		1				2		
	5				3		8	
2				6				5
					4		1	
		2				8		
	8		3					
5				1				6
	2		7				3	
		6				5		2

Figure 2.1. A Sudoku puzzle.

Puzzle 1. *In Figure 2.1, the digits 1–9 appear exactly once in each row, column, and 3 × 3 block. Some of the digits have already been filled in to get you started. Can you complete the remainder of the grid?*

Venues that publish Sudoku typically reassure readers that, the presence of numbers notwithstanding, they involve no mathematics. Representative is the following quotation, written by computer scientist Jean-Paul Delahaye in an article for *Scientific American*:

> Ironically, despite being a game of numbers, Sudoku demands not an iota of mathematics of its solvers. In fact, no operation—including addition or multiplication—helps in completing the grid, which in theory could be filled with any set of nine different symbols (letters, colors, icons, and so on). (Delahaye 2006, 81)

Heaven forbid people should think they are actually doing mathematics while passing the time on a long train ride. Of course, mathematicians will find it a mighty big leap to pass from "no operation helps," to "not an iota of mathematics." *That* only makes sense if you think mathematics is just equivalent to arithmetic.

Sudoku puzzles are exercises in deductive logic. The starting clues constitute the given information, and the solver must deduce their logical consequences. If you have ever worked such a puzzle, you know just how satisfying it is to notice that a certain digit must reside in a specific cell.

We could also mention the success of the game Mastermind as an example of people secretly enjoying the pleasures of logical reasoning. The game works like this: A codemaker and a codebreaker play this game. The codemaker chooses four colored pegs, from among six possible colors, and arranges them in a sequence that is concealed from the view of the

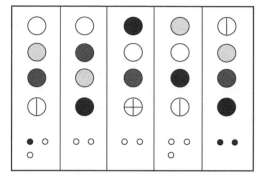

Figure 2.2. A game of Mastermind. The original game uses colored pegs.

codebreaker. The breaker now has ten tries to guess the sequence. After each try, the maker provides information to the breaker regarding how close she is to the correct answer. Specifically, the maker tells the breaker how many of her pegs represent the correct color in the correct place, and also how many of her pegs represent the correct color in an incorrect place. After several tries, the breaker will have amassed a considerable amount of information from which to deduce the answer.

That might be confusing if you have never seen this game before, so here is a puzzle to clarify things:

Puzzle 2. *After five guesses in a game of Mastermind, the codebreaker confronts the situation shown in Figure 2.2. Under each guess, black pegs indicate a correct color in a correct place, while white pegs indicate a correct color in an incorrect place. You can assume that the code contains no repeated colors, and that the six colors you see on the board are the only ones used in the game. Can you work out the code from the information provided?*

To get you started, consider the white peg at the top of the left column. You might wonder if the black peg at the bottom of the column indicates that this peg is in the correct place. But then you notice that in the next column, there is a white peg in the same place, but there is no black peg down below. This tells you that the black peg at the bottom of the first column cannot refer to the white peg at the top. Now keep going!

The game of Mastermind played a significant role in my own interest in logic puzzles. My father taught me to play when I was about eight, but I mostly found the game too difficult at that age. I quickly figured out that I could determine which colors were used in the code by using just one type in each of my first six tries (all blue in the first, all red in the second, and so on), but this only left me four tries to determine which of the 24 orderings was correct. However, when I was the maker and my father the breaker, he *always* found the correct sequence, seemingly without effort and usually with several tries to spare. This remained true even when we tried the advanced version, where a blank space was used as a seventh color. After each success, I sat in rapt attention as he explained his meticulous reasoning. I have a clear memory of thinking, "I want to be able to do that!"

At any rate, that makes two commercially viable ventures that are based on the premise that logic can be fun. It would seem the tedious symbolism of the textbooks is not the whole story.

2.2 Some Classic Logic Puzzles

Among mathematical recreations generally, logic puzzles are the most ecumenical. All that is needed is a bit of patience and some clear thinking. No algebra, geometry, computational skill, or any other sort of specialized knowledge is required. The payoff for solving one, however, is the smile you wear on your face for the rest of the day.

Logic puzzles come in a variety of forms. Here are a few for you to solve. (Solutions to these puzzles appear at the end of the chapter.)

The first three are fairly straightforward, so give them a try.

Puzzle 3. *A boy and a girl are sitting on the front steps of their commune. "I'm a boy," said the one with black hair. "I'm a girl," said the one with red hair. If at least one of them is lying, then which one is which? (Gardner 2006, 36)*

Puzzle 4. *"Feemster owns more than a thousand books," said Albert. "He does not," said George. "He owns fewer than that." "Surely he owns at least one book," said Henrietta. If only one statement is true, then how many books does Feemster own? (Gardner 2006, 36)*

Puzzle 5. *In front of you are three boxes, the first labeled "apples," the second "oranges," and the third "apples and oranges." One box contains apples, one contains oranges, and the other contains apples and oranges. Each label, however, is on the wrong box. Your job is to correctly reassign the labels. You can't see (or smell) what's in any of the boxes. But you are allowed to stick your hand in one of them and remove a single piece of fruit. Which box do you choose, and once you see that piece of fruit, how do you deduce the correct contents of all the boxes? (Bellos 2017, 31)*

This next one is especially famous. There are many variations on the basic idea available in the literature.

Puzzle 6. *Alberta and Bernadette are mucking about in the garden. They come inside. The sisters can see each other's faces, but not their own. Their father, who can see both girls, tells them that at least one of them has a muddy face. He then asks them to stand with their backs to the wall. "Please step forward if you know you have a muddy face," he says. Nothing happens. "Please step forward if you have a muddy face," he repeats. What happens and why? (Bellos 2017, 33)*

The charm of this puzzle is that it seems like the children learn nothing from the man's assertion that at least one of them has a dirty face. Each child already knew that, since she can see that her sister has a dirty face. The key observation is that the father's statement *does* provide information that the children did not already have. They do not learn anything directly about who does and does not have a dirty face, but they do learn something about what the other one knows.

This next puzzle is another that has special meaning for me, since it appeared in a book my parents gave me when I was very young. It was called *The Great Book of Puzzles,* and it was printed on the sort of paper you were expected to write on. Many of the puzzles were suitable for young children, but it also contained a difficult logic puzzle. A version of this puzzle was originally published in *Life* magazine in 1962, but it has been republished many times in various forms.

Puzzle 7. *Five different people live in five different houses. These houses are arranged in a line. Each house is a different color: red, green, yellow, blue, or ivory. Each of the five people comes from a different country: England, Spain, Norway, Japan, or Ukraine. Each owns a different pet: a dog, a fox, a horse, a monkey, or a snail. Each prefers a different beverage: water, tea, milk, coffee, or orange juice. And each practices a different profession: There is a teacher, an accountant, a daredevil, a lawyer, and a doctor.*

When we speak of one house being to the left or the right of the other, we are referring to your left and your right as you are facing the houses. The "first" house then refers to the one farthest to your left. You are given the following facts:

1. *The Englishman lives in the red house.*
2. *The Spaniard owns the dog.*
3. *Coffee is drunk in the green house.*
4. *The Ukrainian drinks tea.*
5. *The green house is immediately to the right of the ivory house.*
6. *The daredevil owns a snail.*

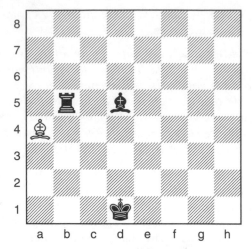

Figure 2.3. The chess position for puzzle 8.

7. *The teacher lives in the yellow house.*
8. *Milk is drunk in the middle house.*
9. *The Norwegian lives in the first house.*
10. *The accountant lives next to the house with the fox.*
11. *The teacher lives next to the house where the horse is kept.*
12. *The lawyer drinks orange juice.*
13. *The doctor is Japanese.*
14. *The Norwegian lives next to the blue house.*

Who owns the monkey?

Of course, as a kid, this was entirely beyond me. I was nonetheless intrigued, mostly because the question "Who owns the monkey?", printed in a large font at the bottom of the page, struck me as inherently funny. As an amusing aside, let me mention that the version of this puzzle that appeared in the book my parents gave me (the one for children, recall) did not say that the five people enjoyed different beverages. It said they smoked five different brands of cigarettes.

Finally, let me close with a logic puzzle that is of a different sort altogether. It requires a knowledge of the rules of chess, though I would hasten to add that no skill as a player is necessary. It was devised by Raymond Smullyan.

Puzzle 8. *In the position shown in figure 2.3, the white king has been removed from the board. On what square must he stand, if we assume this position was reached during a game of chess in which only legal moves* *were played? You may assume that white is moving up the board, while black is moving down. (Smullyan 1981, 3)*

2.3 Puzzles in Propositional Logic

While the examples thus far could all be fairly described as logic puzzles, they are mostly not the sort of puzzles with which we shall be concerned in this book. Instead, we will be concerned with what might loosely be called puzzles in propositional logic.

Let us warm up with a few examples of what I have in mind. Our first two come from Lewis Carroll, whose work will form the basis of Part II. Each puzzle involves deducing the consequences of lists of categorical statements. You will notice that in each case, there are two terms that appear exactly once among the premises, while all the other terms appear twice. Your goal is to draw a conclusion that involves only the two terms that appear once.

Puzzle 9. *What can you conclude from the following statements?*

- *My saucepans are the only things I have that are made of tin.*
- *I find all your presents very useful.*
- *None of my saucepans are of the slightest use.*

(Carroll 1958, 112)

Puzzle 10. *Let us try a harder one. What can you conclude this time?*

- *No kitten that loves fish is unteachable.*
- *No kitten without a tail will play with a gorilla.*
- *Kittens with whiskers always love fish.*
- *No teachable kitten has green eyes.*
- *No kittens have tails unless they have whiskers.*

(Carroll 1958, 119)

Part III is devoted largely to the work of Raymond Smullyan. He was especially adept with puzzles involving knights and knaves, in which we assume that knights only make true statements, while knaves only make false statements. The two tribes are visually indistinguishable, meaning that the only way to get information from them is to work out the logical consequences of their often cryptic statements.

To give you an idea of how this works, here is a simple example whose solution we shall discuss below.

Puzzle 11. *You meet two people, who are referred to as A and B. A says, "At least one of us is a knave." What can you determine about the two people? (Smullyan 1978, 20)*

To solve this, we reason that if A is a knave, then it follows that his statement that there is at least one knave among them is true. This is a contradiction, since knaves always make false statements. We conclude that A is a knight, and that his statement must be true. Since there really is at least one knave among them, that person must be B. So the solution is that A is a knight and B is a knave.

Puzzles about liars and truthtellers long predate Raymond Smullyan's work. Here is an especially famous example, first presented by philosopher Nelson Goodman in 1931.

Puzzle 12. *All the men of a certain country are either nobles or hunters, and no one is both a noble and a hunter. The male inhabitants are so nearly alike that it is difficult to tell them apart, but there is one difference: Nobles never lie, and hunters never tell the truth.*

Three of the men meet one day and Ahmed, the first, says something. He says either, "I am a noble," or "I am a hunter." (We don't know yet which he said.)

Ali, the second man, heard what Ahmed said, and in reply to a query, answered, "Ahmed said, 'I am a hunter.'" Then Ali went on to say, "Azab is a hunter."

Azab was the third man. He said, "Ahmed is a noble."

Now the problem is, which is each? How do you know? (Goodman 1972, 451)

Hopefully, this has whetted your appetite for what is to come. I often tell my students that math is easy; it is math classes that are hard. Perhaps I can persuade you of something similar with the remainder of this book. Textbooks of formal logic do not make for light reading, but a nice logic puzzle can be very nice indeed.

As we will see, the formalities and the puzzles are two sides of the same coin, and they are not as distinct as you might think.

2.4 Notes and Further Reading

A rich mathematical theory underpins the games of Sudoku and Mastermind. The book by Rosenhouse and Taalman (2012) is a helpful resource for Sudoku, while Bewersdorff (2004, ch. 32) will give you a good start on Mastermind. The books by Martin Gardner (2006) and Alex Bellos (2017), both mentioned in Section 2.2, are especially masterful collections of puzzles. They should reside on the shelf of anyone interested in recreational mathematics and logic.

The Sudoku puzzle shown in Figure 2.1 was created by Laura Taalman and Phil Riley. Their company, Brainfreeze Puzzles, has produced many fascinating books of original Sudoku puzzles.

Puzzle 6 is often referred to as the "muddy children puzzle." It is the simplest representative of a genre known as puzzles in epistemic logic. Space restrictions made it impossible to discuss this topic in more detail, but the interested reader can have a look at the book by van Ditmarsch and Kooi (2015).

Puzzle 8 is an example of a *retrograde analysis chess problem*. In addition to his more traditional books of logic puzzles, Raymond Smullyan produced two books of such puzzles (Smullyan 1979, 1981). I had originally intended to include a more thorough discussion of such puzzles, but ultimately decided they were too esoteric. However, Puzzle 8 is an especially famous representative of the genre, and I suspect that many readers will possess the necessary background knowledge to enjoy it.

2.5 Solutions

Puzzle 1. The solution is shown in Figure 2.4.

3	9	1	5	4	8	2	6	7
6	5	7	9	2	3	4	8	1
2	4	8	1	6	7	3	9	5
9	6	5	2	8	4	7	1	3
7	3	2	6	9	1	8	5	4
1	8	4	3	7	5	6	2	9
5	7	3	8	1	2	9	4	6
4	2	9	7	5	6	1	3	8
8	1	6	4	3	9	5	7	2

Figure 2.4. The completion of the Sudoku square from Puzzle 1.

Puzzle 2. The solution is shown at the far right of Figure 2.5. One way to see that this must be correct is as follows: There is one black peg under the first guess. Which of the four disks does it represent? Not the white, as then there would be a black peg under the second guess. Likewise, it does not represent the dark-gray or the lined disk, since then there would

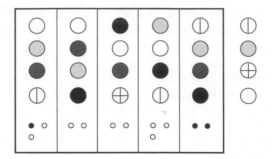

Figure 2.5. The far right column represents the solution to Puzzle 2. To make the explanation easier to follow, the original puzzle is also shown here.

be black pegs underneath the third and fourth guesses, respectively. We conclude that it represents the light-gray disk.

There are two black pegs under the fifth guess. One of them is for the light-gray disk. The other one cannot be the black disk, since then a black peg would appear under the second guess as well. We have already seen that the dark-gray disk is not correct. Therefore, the second black peg is for the lined disk at the top.

Moreover, since there are two black pegs underneath the fifth guess, representing the lined and light-gray disks, we conclude that the dark-gray and black disks in the fifth guess are simply wrong. That is, these colors do not appear in the solution. It follows that the other two colors in the solution are white and crossed. The crossed disk cannot be at the bottom, since then there would have been a black peg under the third guess. That puts the crossed disk in second place from the bottom, forcing the white disk to go to the bottom.

Puzzle 3. We are told there is one boy and one girl on the steps. If exactly one is lying, then the other is telling the truth. This would imply that there are either two boys or two girls on the step, depending on which one is lying, and this is a contradiction. The only other possibility is that both children are lying. Thus, the one with black hair is the girl, and the one with red hair is the boy.

Puzzle 4. Since the statements made by Albert and George contradict each other, it follows that one is true and the other is false. Since we are told there is only one true statement among the three, we conclude that Henrietta's statement must be false. Thus, Feemster does not actually own any books.

Puzzle 5. Reach into the box labeled "apples and oranges." We know this sign is misplaced. Therefore, if we remove an apple, then we know the correct label for this box is "apples." If we remove an orange, then the correct label is "oranges." For simplicity, let us assume that the correct label for this box is "apples." Thus, we remove the sign from the box originally labeled "apples" and move it to the box originally labeled "apples and oranges." Now, to which box should we move the sign "apples and oranges"? If we move it to the box originally labeled "apples," then we would be forced to leave the "oranges" sign precisely where it is. But this does not work, since we were told that every sign begins on the wrong box. So the only possibility is to move the "oranges" sign to the box originally labeled "apples" and to move the "apples and oranges" sign to the drawer originally labeled "oranges."

Puzzle 6. Each sister knows that at least one of them has a smudge on her face, but she can only see the other sister's face. Now, if either sister sees that the other has a clean face, then she will know immediately that her own face must be dirty and will therefore step forward at her father's request. Since neither one did step forward, it must be that each one sees that the other has a dirty face. But by the time their father makes his request for the second time, each will have gone through precisely this reasoning. Thus, they will both step forward when their father makes his request for the second time.

Puzzle 7. The full solution is shown in Table 2.1.

Here is one method for reasoning your way to this solution. We are given that the Norwegian is in the first house (9). Therefore, the second house is blue (14). The first house is therefore not red (1), or blue (14), or green or ivory (5). Therefore, the first house is yellow. It then follows that the Norwegian is the teacher (7) and that the horse lives in the second house (11).

We next investigate the remaining colors. Fact 5 coupled with the fact that yellow and blue are accounted for tells us that the third, fourth, and fifth houses are either red, ivory, green or ivory, green, red.

But ivory, green, red leads to a contradiction as follows: If the fifth house is red, then the Englishman lives in the fifth house (1). We also know that coffee is drunk in the fourth house (3), and milk is drunk in the third house (8). Since we already know the Norwegian and the Englishman are in the first and fifth houses, and since the Ukrainian drinks tea (4), he must be in the second house. Since we know that the teacher lives in the first house, and that tea, milk, and coffee are drunk in the second through fourth houses, we are forced to put the lawyer and orange juice in the fifth house (12). We also know that the Spaniard and the dog go together (2), as do the daredevil and the snail (6). These pairs must go in houses three and

TABLE 2.1.
The solution to the monkey puzzle.

Norway	Ukraine	England	Spain	Japan
Water	Tea	Milk	Orange juice	Coffee
Yellow	Blue	Red	Ivory	Green
Teacher	Accountant	Daredevil	Lawyer	Doctor
Fox	Horse	Snail	Dog	Monkey

four, in some order. But the Japanese person must also be in one of these two houses. This forces us to conclude that the Japanese person is the daredevil, which is a contradiction (13). Therefore, the correct sequence is red, ivory, green.

We now immediately obtain that the coffee drinker is in the fifth house (3) and that the Englishman lives in the third house (1).

What does the Norwegian drink? Not milk or coffee, since they are accounted for. Not orange juice, since the lawyer (not the teacher) drinks that (12). And not tea, since that is the Ukranian's drink (4). Therefore, the Norwegian drinks water.

The Spaniard must live in the fourth or fifth house, since the first and third contain the Norwegian and the Englishman, and the second contains the horse (while the Spaniard has a dog (2).) By fact (4), we know that the Ukrainian drinks tea. Given what we already know, that puts both of them (the Ukranian and tea) either in the second or the fourth house. Suppose they are in the fourth house. Then the Spaniard is in the fifth, and by elimination, the Japanese person is in the second. However, by elimination, we also have that orange juice is drunk in the second house. This is impossible, since the Japanese person is the doctor (13), while it is the lawyer who drinks orange juice (12). We have reached a contradiction.

Now the puzzle unravels. To avoid the contradiction in the last paragraph, we must place the Ukranian and the tea drinker in the second house. Elimination now puts the orange juice in the fourth house. This puts the lawyer in the fourth house as well (12). Since the Japanese person is the doctor, the only place left for him is the fifth house. Elimination then puts the Spaniard in the fourth house, along with the dog (2). We know that the daredevil and the snail go together (6), and the only place left for them is the third house. Elimination now puts the accountant in the second house. Since the accountant lives next to the fox (10), we must put the fox in the first house.

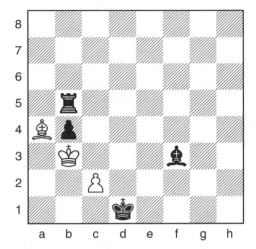

Figure 2.6. The position two moves prior to the one given in figure 2.3.

That accounts for everything except the monkey. The only place left for him is the fifth house, with the Japanese person.

Puzzle 8. The white king either stands on b3, or black is in check from the white bishop on a4. (Recall that two kings can never stand next to each other, implying that the white king cannot stand on c2.) Now, a white king standing on b3 would be in check from black's rook and bishop simultaneously. Black has no last move that could have delivered such a check, so this scenario is impossible.

It follows that black is currently in check. Since the white bishop has no last move that could have delivered such a check, we conclude that the white king has just moved from b3, thereby unblocking the bishop. But now we again face the problem of explaining how the white king could be in check simultaneously from the black rook and bishop. The only possibility is that two moves previously, the position was roughly as shown in Figure 2.6.

Black then moved his bishop to d5, giving check. White blocked the check by moving his pawn from c2 to c4. Black now captured en passant, removing both pawns from the board and leaving white in check from the rook and bishop simultaneously. White then captured the black pawn now on c3 with his king. So the white king now stands on c3.

Puzzle 9. If all of *your* presents are useful, but none of my saucepans are useful, then it must be that you did not give me any of my saucepans as presents. But if none of your presents were among my saucepans, and my saucepans are the only things I have that are made of tin, then it must be that none of *your* presents were made of tin.

Puzzle 10. Begin by scanning the five statements, looking for the two categories that only appear once each. We find that our conclusion should relate having green eyes with a willingness to play with a gorilla. It will also be helpful to rewrite the second statement in the equivalent form, "All kittens that play with gorillas have tails," and the fifth statement in the equivalent form, "All kittens with tails have whiskers." Now, here is one way to obtain a conclusion of the desired form: From the second and fifth statements, conclude that all kittens who play with gorillas have whiskers. Combine this with the third statement to get that all kittens that play with gorillas love fish. Combine *this* with the first statement to get that all kittens that play with gorillas are teachable. Finally, combine this with the fourth statement to conclude that no kittens that play with gorillas have green eyes. (This is equivalent to saying that no kittens with green eyes will play with gorillas, so this is also an acceptable answer.)

Puzzle 12. The key observation is that no one can claim to be a hunter. A hunter who made that claim would be telling the truth, while a noble who made that claim would be lying. We would get a contradiction either way. We conclude that Ali lied when he claimed that Ahmed said he was a hunter, which implies that Ali actually *is* a hunter. Since Ali is a hunter, his claim that Azab is a hunter is false. Therefore, Azab is a noble. It follows that his statement that Ahmed is a noble is true. Thus, the solution is that Ahmed is a noble, Ali is a hunter, and Azab is a noble.

PART II
Lewis Carroll and Aristotelian Logic

CHAPTER 3

Aristotle's Syllogistic

Lewis Carroll pioneered the field of recreational logic by publishing, in 1886, a short book called *The Game of Logic*. Noticing that logic could be pursued just for fun was no small accomplishment, given the scholarly climate at the time he was writing. The logic of his day was still dominated by the traditional logic of Aristotle and his followers, and this was primarily what was taught to university students. We cannot appreciate Carroll's work until we have a solid grounding in Aristotelian logic.

Aristotle wrote six works related to logic. Today these works are known collectively as the *Organon*. Only one of these works was devoted primarily to issues in formal logic, and that work will be our focus here. Section 8.1 has some discussion of the *Organon* as a whole.

3.1 The Beginning of Formal Logic

The Greek philosopher Aristotle inaugurated the study of formal logic sometime during the 300s BCE, in a work referred to as the *Prior Analytics*.

For centuries prior to Aristotle, there were philosophers who not only pondered deep questions but also wrote down their thoughts. One wonders, then, why none of *them* felt it worthwhile to undertake a formal study of the principles of correct reasoning. Certainly the work of Plato and other pre-Aristotelian philosophers included intimations of logical ideas, but they made no systematic attempt to study logic in its own right (Kneale and Kneale 1962, 1–22).

A plausible answer is that Aristotle wrote at a time when Greek democracy still represented a relatively novel form of government. For the first time, political change could only be affected by persuading others to go along, and this put a premium on skill in rhetoric and argumentation. However, arguments can be persuasive without being correct. Skillful debaters can appeal to people's emotions or biases, as opposed to their reason. When such arguments can direct the course of society, the development of methods for clear thinking becomes imperative (Shenefelt and White 2013, 33–48).

We can only speculate about Aristotle's motives, however, since he does not pause to tell us why he chose to address this topic. He just dives right in. The *Prior Analytics* opens with:

> We must first state the subject of our inquiry and the faculty to which it belongs: its subject is demonstration and the faculty that carries it out demonstrative science. We must next define a premise, a term, and a syllogism, and the nature of a perfect and of an imperfect syllogism; and after that, the inclusion or noninclusion of one term in another as in a whole, and what we mean by predicating one term of all, or none, of another.
>
> A premise, then, is a sentence affirming or denying one thing of another. This is either universal or particular or indefinite. By universal I mean the statement that something belongs to all or none of something else; by particular that it belongs to some or not to some or not to all; by indefinite that it does or does not belong, without any mark to show whether it is universal or particular. (Aristotle 2001, 65)

It seems that Aristotle also inaugurated the tradition that logic texts should be turgid and dull.

The *Prior Analytics* addressed a particular kind of argument, referred to as a "syllogism." Aristotle writes: "A syllogism is a discourse in which, certain things being stated, something other than what is stated follows of necessity from their being so" (Aristotle 2001, 66). Deducing the correct conclusion of a syllogism can be seen as bringing out and making explicit information that was contained in disguised form in the premises.

Aristotle noticed that some arguments are made valid purely by their form, and not because of the empirical content of their premises. Hence the term "formal logic." An example is the argument with which I opened the book:

All cats are mammals.

All mammals are animals.

All cats are animals.

The conclusion follows from the premises, but not because of anything we know about cats, mammals, or animals. Rather, it is valid because it is a concrete instance of the abstract form:

> All Ss are M.
>
> All Ms are P.
> _____
> All Ss are P.

and this form is immediately recognized as valid. Let us use the letters S and P to distinguish the subject of a proposition from its predicate. The letter M denotes the middle term, as we shall discuss momentarily.

Let us now consider the following question: If you are given two statements of this form involving the terms S, M, and P with M appearing in both statements, what conclusion, if any, can be drawn relating S and P?

Sometimes this is an easy question, as in the example just noted. Here is another example where the conclusion is very natural:

> All Ss are M.
>
> No Ms are P.
> _____
> No Ss are P.

In both of these examples, the conclusion follows very naturally from understanding the meanings of the words involved.

Other times, however, it is less simple:

> No Ss are M.
>
> All Ps are M.
> _____

If you are unaccustomed to this sort of thing, then it might take a moment to see what follows. Matters become more straightforward if you recognize that the first statement is equivalent to, "No Ms are S." If you then reorder the statements you have

> All Ps are M.
>
> No Ms are S.
> _____

We recognize this as another instance of the form we previously considered, and quickly deduce the conclusion "No Ps are S." This is equivalent to "No Ss are P."

We could also have reached this conclusion by employing a Venn diagram, as shown in Figure 3.1.

In still other cases, no conclusion of the desired sort is possible:

> No Ss are M.
>
> No Ps are M.
> _____

There is nothing to be said here regarding the relationship of S to P.

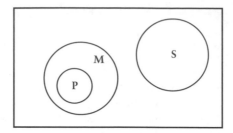

Figure 3.1. A Venn diagram representing the premises "No Ss are M" and "All Ps are M." The correctness of the conclusion "No Ss are P" is now readily apparent.

Since there are only so many pairs of statements of the required form, we might hope to provide a complete classification of the valid inferences that can be drawn. Such a classification is a principle accomplishment of the *Prior Analytics*.

Finding such a classification is tedious but does not seem very difficult. Working out the deductive consequences of pairs of premises requires little more than understanding the words and a moment's calm reflection. Drawing a diagram will help with any difficult cases.

Nonetheless, high-brow textbook treatments of Aristotelian logic are laden with shocking amounts of jargon and notation, much of which serves only to make simple ideas seem complex. Moreover, for centuries after Aristotle, the study of formal logic was dominated by picayune concerns over the nature of terms and propositions and their arrangements into syllogisms. There is something about the density of Aristotle's prose that has provoked numerous scholars over the centuries, especially in the Middle Ages, to produce dense, unreadable tomes of their own. Let us, then, devote a few sections to considering the horror, if only to better appreciate, in the next chapter, Lewis Carroll's far more enjoyable approach to the same material.

As you let this material wash over you (and I promise there will not be a test at the end of it), ponder the situation of the medieval university students who were expected to master all of this.

3.2 Proposition Jargon

We start with the premises themselves, which are said to be *categorical propositions*. By this it is meant that they involve relationships among categories of things. For example, "All cats are mammals" puts forth a relationship between the category "cats" and the category "mammals."

The categories are typically referred to as *terms*, and Aristotelian logic as a whole is sometimes called *term logic*. This should be contrasted with *propositional logic*, which we shall discuss in Part III. If you think of a term as being equivalent to what mathematicians refer to as a set (which is to say, an arbitrary collection of objects), then you will not go wrong with respect to anything we shall discuss.

In logic, an argument with two premises and a conclusion is called a *syllogism*, from Greek words translating roughly as "with reason." If the premises and conclusion of a syllogism are categorical propositions, then we have a *categorical syllogism*. Aristotle, then, was the first to classify categorical syllogisms.

Categorical propositions come in four basic forms:

1. All Ss are P.
2. No Ss are P.
3. Some Ss are P.
4. Some Ss are not P.

Such propositions have four parts. The term playing the role of S is the *subject* of the proposition, while the term playing the role of P is the *predicate*. The form of the verb "to be" that connects the subject and the predicate is called the *copula*, from Latin words translating as "to fasten together" or "to connect." The copula connects the subject to the predicate. The opening word—all, some, or no—is the *quantifier*.

Thus, in the proposition "All Ss are P," the quantifier is "All," the subject and predicate are S and P, respectively, and "are" is the copula. This proposition affirms an attribute of everything in category S; specifically, that everything in S is also in P. Statements of this sort are therefore described as *universal affirmative*. For similar reasons, the other three forms in our list are described as *universal negative*, *particular affirmative*, and *particular negative*, respectively.

We take for granted that all of our discussions take place in a clearly defined *universe of discourse*. That is, we are generally not interested in reasoning about all objects simultaneously, but only about some small subset of them. In talking about cats, mammals, and animals, for example, we might take the universe of discourse to be "living things." It is with respect to this universe that the quantifier of a categorical proposition is to be understood. To say "All Ss are P" or "No Ss are P" is to say, respectively, that every S or no S in the universe is P. However, the statement "Some Ss are P" is understood to mean that at least one S in the universe is P. In everyday conversation, a statement beginning with "some" usually implies "more than one," but that is not the convention used in Aristotelian logic.

TABLE 3.1.
Abbreviations for the types of categorical propositions.

Abbreviation	Type of sentence	Example
A	Universal affirmative	All Ss are P.
E	Universal negative	No Ss are P.
I	Particular affirmative	Some Ss are P.
O	Particular negative	Some Ss are not P.

In a universe of discourse, we might wish to discuss the set of objects that lack P. This set is called the *complement* of P and is denoted by "non-P."

The Latin word "affirmo" means "I affirm." Since the first two vowels in this word are *A* and *I*, it has become standard to refer to universal affirmatives and particular affirmatives as *A* statements and *I* statements, respectively. Likewise, the Latin word "nego" means "I deny." Considering the vowels in this word leads to universal negatives and particular negatives being referred to as *E* and *O* statements, respectively. These definitions are summarized in Table 3.1.

We see that there are two important questions when considering a proposition: Is it universal or particular, and is it affirmative or negative? The former attribute is the proposition's *quantity*, while the latter is its *quality*. Thus, we can say that *A* and *I* statements are affirmative in quality, while *E* and *O* statements are negative.

Looking anew at two of our previous examples:

(1) All Ss are M.   (2) All Ss are M.

All Ms are P.   No Ms are P.
_____

All Ss are P.   No Ss are P.

we find that (1) involves three statements that are universal in quantity and affirmative in quality. Argument (2) likewise involves three statements that are universal in quantity. However, in (2), only the first premise is affirmative in quality, while the other premise and the conclusion are negative.

Moving on, the term that appears in both premises of a categorical syllogism is called the *middle term*, while the other two are said to be the *extreme terms*. The predicate of the conclusion is called the *major term*, while the subject of the conclusion is the *minor term*. The premise containing the major term is the *major premise*, while the premise containing the minor term is the *minor premise*. In expressing our syllogisms, it is customary to place the major premise first and the minor premise second.

This terminology seems apt when the middle term is the subject of one premise and the predicate of the other. In the example

> All mammals are animals.
>
> All cats are mammals.
> _____
> All cats are animals.

we have a progression of terms from larger to smaller: from animals to mammals to cats. It would seem that "mammals" does indeed appear in the middle, and that "animals," by virtue of being the largest class, is plausibly called the major term.

However, in examples such as this:

> (3) All Ps are M.
>
> No Ss are M.
> _____
> No Ss are P.

the language of middle, major, and minor terms seems less appropriate. However, we have seen that reversing the order of the premises, while replacing "No Ss are M" with the equivalent "No Ms are S," reduced the argument to the same form as (2), which perhaps justifies using this terminology after all.

3.3 Operations on Propositions

We have seen that it is sometimes useful to replace a categorical proposition with one that is logically equivalent. Let us investigate this further.

Given a categorical proposition, the proposition identical in quality and quantity, but with the subject and predicate reversed, is called the *converse* of the proposition. For example, the converse of "All Ss are P" is "All Ps are S," while the converse of "No Ss are P" is "No Ps are S."

Plainly, "All cats are mammals" is not equivalent to "All mammals are cats." This realization can be expressed by saying, "*A* statements do not convert." The same is true for *O* statements: "Some dogs are not terriers" is not equivalent to "Some terriers are not dogs." Conversion works better for *E* and *I* statements: "No cats are dogs" is equivalent to "No dogs are cats," and "Some cats are orange" is equivalent to "Some orange things are cats." Thus, *E* and *O* statements convert, while *A* and *I* statements do not.

Most would accept these conversions as obviously correct, but Aristotle sought to justify them nonetheless:

> First, then, take a universal negative with the terms A and B. If no B is A, neither can any A be B. For if some A (say C) were B, it would not be true that no B is A; for C is a B. But if every B is A then some A is B. For if no A were B, then no B could be A. But we assumed that every B is A. Similarly, too, if the premise is particular. For if some B is A, then some of the As must be B. For if none were, then no B would be A. But if some B is not A, there is no necessity that some of the As should not be B; e.g. let B stand for animal and A for man. Not every animal is a man; but every man is an animal. (Aristotle 2001, 66)

Has prose such as this ever allayed the doubts of anyone skeptical of the conversions?

In addition to conversions, Aristotle also made use of a certain *immediate inference*, by which is meant an inference based on one premise. Given "All Ss are P," Aristotle concluded that "Some Ps are S." This is acceptable if we assume that *A* statements have *existential import*. A proposition is said to have existential import if it entails the existence of at least one object. If we construe "All Ss are P" to be equivalent to "At least one S exists, and all Ss are P," then it is fine to conclude, "Some Ps are S."

Since Aristotle's investigations were primarily motivated by practical concerns, it is understandable that he would assume his terms were not empty. Why would practical people be discussing the properties of nonexistent things? Moreover, in everyday discourse, it seems reasonable to assume that *I* and *O* statements have existential import. When you assert that "Some Ss are P," the implication is that you are partitioning the class of Ss into those that do or do not belong to P. If there are no Ss, then it seems like an abuse of language to hold forth on properties possessed by only some of them.

However, modern logicians are unanimous in finding the assumption of existential import undesirable for *A* and *E* statements. It seems perfectly reasonable to say "All unicorns are horses" without being committed to the existence of unicorns. For mathematicians, this is a very natural point, since we routinely deduce properties of objects that are not known to exist. That is, we might define some abstract object and then say: "We do not know if any such objects exist, but if they do, then they must have certain other properties."

Let us consider some operations other than conversion that we might perform on propositions. If we change the quality of a proposition, and then take the complement of the predicate, the result is the *obverse* of what we started with. Consider "All Ss are P." Changing the quality from affirmative to negative while negating P leads to "No Ss are non-P."

Likewise, the obverse of "No Ss are P" is "All Ss are non-P." The obverse of "Some Ss are P" is the somewhat awkward "Some Ss are not non-P." For O statements, negating the predicate and changing the quality effectively cancel each other out. The obverse of "Some Ss are not P" is "Some Ss are non-P." In each case, the obverse is logically equivalent to the original statement.

Our final operation is *contraposition*. The contrapositive of a proposition is found by switching the roles of S and P, and then taking the complement of each. The resulting proposition is the *contrapositive* of what you started with. The contrapositive of "All Ss are P" is "All non-Ps are non-S." The contrapositive of "No Ss are P" is "No non-Ps are non-S." The contrapositives of "Some Ss are P" and "Some Ss are not P" are, respectively, "Some non-Ps are non-S" and "Some non-P are not non-S."

Let us put all of this together:

- Start with "All Ss are P."
- This obverts to the logically equivalent "No Ss are non-P."
- This then converts to the equivalent "No non-Ps are S."
- Obverting once more leads to "All non-Ps are non-S," which is the contrapositive of "All Ss are P."

We have shown that an *A* statement is equivalent to its contrapositive.

Of course, a concrete example makes the point with greater force: If all cats are mammals, then all non-mammals are non-cats.

If you care to work out the details, you will find that *A* and *O* statements are equivalent to their contrapositives, while *E* and *I* statements are not. (From "No cats are dogs," it certainly does not follow that "No non-dogs are non-cats.") Our findings about conversion, obversion, and contraposition are recorded in Table 3.2.

The relationships among a statement and its converse, obverse, and contrapositive form the basis for the traditional *Square of Opposition*, shown in Figure 3.2.

The terminology in the square in Figure 3.2 is explained as follows:

- Two propositions are *contradictory* if it is impossible either for both to be true, or both to be false. In other words, one is true and the other is false. "All Ss are P" and "Some Ss are not P" are contradictories.
- Two propositions are *contraries* if they cannot both be true, but they can both be false. "All Ss are P" and "No Ss are P" are contraries.
- Two propositions are *subcontraries* if they cannot both be false, but they can both be true. "Some Ss are P" and "Some Ss are not P" are subcontraries.
- A proposition Y is a *subaltern* of a proposition X if Y must be true whenever X is true, and X must be false whenever Y is false. "Some

TABLE 3.2.
Three operations on categorical propositions.

Operation	Effect		Equivalent?
Conversion	A	All Ps are S	No
	E	No Ps are S	Yes
	I	Some Ps are S	Yes
	O	Some Ps are not S	No
Obversion	A	No Ss are non-P	Yes
	E	All Ss are non-P	Yes
	I	Some Ss are not non-P	Yes
	O	Some Ss are non-P	Yes
Contraposition	A	All non-Ps are non-S	Yes
	E	No non-Ps are non-S	No
	I	Some non-Ps are non-S	No
	O	Some non-Ps are not non-S	Yes

Notes: The second column records the effect of carrying out the operation on the given type of statement. Refer to Table 3.1 for the definitions of *A*, *E*, *I*, and *O* statements.

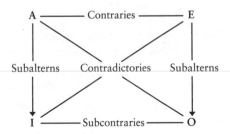

Figure 3.2. The traditional square of opposition.

S is P" is subaltern to "All Ss are P," and "Some Ss are not P" is subaltern to "No Ss are P."

The arrows in Figure 3.2 indicate that the truth of an *A* or *E* statement immediately implies the truth of the corresponding *I* or *O* statement. In calling this the "traditional" square of opposition, we are assuming, with

Aristotle, that all propositions have existential import. That is, empty terms are disallowed.

3.4 Figures and Moods

Had enough? Sorry, but we are just getting started.

You see, having developed a thorough grasp of propositions, we must now see how they might be stacked up into valid syllogisms.

In the premises, there are four possible arrangements of the middle terms with respect to the extremes:

M—P	P—M	M—P	P—M
S—M	S—M	M—S	M—S
S—P	S—P	S—P	S—P

These are referred to as the various *figures* of a categorical syllogism. They are numbered first through fourth, respectively, as recorded in Table 3.3. Aristotle himself only discussed the first three figures, and it has been the source of some scholarly debate as to why he did not discuss the fourth (Rose 1968, Ch. 1). For our purposes, however, it will be simpler to include the fourth figure.

In each figure, an argument is uniquely characterized by the statement types of its propositions. For example, a first-figure argument whose propositions are all of type A must be identical in form to the example given in the first row of Table 3.3. We can abbreviate this by saying the argument is of the form AAA. This triple is called the *mood* of the argument. The moods of the other examples given in Table 3.3, from second to fourth, are EAE, IAI, and EIO, respectively. A syllogism is completely characterized by its figure and mood.

There are three propositions in a syllogism, each of which can be any of the four types A, E, I, O. This makes a total of $4^3 = 64$ moods for each figure. Since there are four figures, there are 256 potential syllogisms. That is not really so many, especially since most of them are obviously invalid. After examining them all, we can readily identify 15 forms that are unquestionably valid.

The valid forms in the first and second figures are shown in Table 3.4. The Latin words attached to each form are mnemonic devices devised by medieval logicians. The sequence of vowels in each word matches the mood of the argument. Thus, if you refer casually to *Ferio*, for example, anyone versed in Aristotelian logic will understand that you mean the first figure argument with mood EIO.

TABLE 3.3.
The four figures into which syllogisms can fall.

Figure	Abstract form	Example
First	M—P	All mammals are animals.
	S—M	All cats are mammals.
	S—P	Therefore, all cats are animals.
Second	P—M	No cats are dogs.
	S—M	All terriers are dogs.
	S—P	Therefore, no terriers are cats.
Third	M—P	Some cats are orange.
	M—S	All cats are mammals.
	S—P	Therefore, some mammals are orange.
Fourth	P—M	No dogs are cats.
	M—S	Some cats are orange.
	S—P	Therefore, some orange things are not dogs.

Note: P denotes the major term, M denotes the middle term, and S denotes the minor term.

TABLE 3.4.
The valid forms in the first figure (left) and second figure (right).

Mood	Mnemonic	Mood	Mnemonic
AAA	Barbara	*EAE*	Cesare
EAE	Celarent	*AEE*	Camestres
AII	Darii	*EIO*	Festino
EIO	Ferio	*AOO*	Baroco

The sample arguments given for the first and second figures in Table 3.3 are instances, respectively, of Barbara and Cesare.

The valid forms in the third and fourth figures are given in Table 3.5. The sample arguments in the third and fourth figures of Table 3.3 are instances, respectively, of Disamis and Fresison.

TABLE 3.5.
The valid forms in the third figure (left) and fourth
figure (right).

Mood	Mnemonic	Mood	Mnemonic
IAI	Disamis	AEE	Camenes
AII	Datisi	IAI	Dimaris
OAO	Bocardo	EIO	Fresison
EIO	Ferison		

It would try patience beyond reason to give examples for all the forms, but here are two more to help cement the ideas. You are encouraged to devise examples of your own to illustrate other figures and moods.

The following is an instance of Festino. Note that the premises express the mood *EIO*, precisely as the sequence of vowels in "Festino" suggests:

> No cats like veterinarians.
>
> Some animals like veterinarians.
> _____
> Some animals are not cats.

Here is an instance of Bocardo, with the premises expressing the mood *OAO*:

> Some dogs are not terriers.
>
> All dogs are quadrupeds.
> _____
> Some quadrupeds are not terriers.

I described these 15 forms as unquestionably valid. There are an additional nine that are only questionably valid. An example is the mood *AAI* in the third figure, which receives the mnemonic Darapti. Here are two arguments of that form:

> All cats are mammals.   All unicorns have horns.
>
> All cats are animals.   All unicorns are horses.
> _____   _____
> Some animals are mammals.   Some horses have horns.

The argument on the left seems valid, but the one on the right does not.

Another example of the same phenomenon occurs again in the mood *AAI*, this time in the fourth figure. This form is known as Bramantip:

All dogs are mammals.   All unicorns are horses.

All mammals are animals.   All horses are mammals.

Some animals are dogs.   Some mammals are unicorns.

Again, the argument on the left is valid, while the argument on the right is not.

The issue in these cases is, again, existential import. "Some horses have horns" is understood to imply that horses with horns exist, but in fact they do not. The problem is that we arrived at our conclusion by taking the premises to entail that unicorns are examples of horned horses, but unicorns do not exist. Thus, it would seem that Darapti is a valid form only if the middle term is not empty. Bramantip illustrated a similar pathology, except it was the emptiness of the predicate term that caused the problem.

If we include among our valid syllogisms those that require an assumption of at least one nonempty term, then we obtain the four forms shown in Table 3.6.

Are there any forms that require the assumption that S is nonempty? Indeed there are. An example is the mood EAO in the first figure:

No Ms are P.   No horses are fish.

All Ss are M.   All unicorns are horses.

Some S is not P.   Some unicorns are not fish.

The argument form on the left is valid when S is not empty. Without this assumption, you encounter the problem shown by the argument on the right.

However, there is another problem: The conclusion "Some S is not P" is not the strongest that can be drawn from the premises. We could instead

TABLE 3.6.
Forms that are valid in the presence of an assumption regarding the nonemptiness of one of their terms.

Figure	Mood	Mnemonic	Assumption
Third	AAI	Darapti	M is not empty
	EAO	Felapton	M is not empty
Fourth	AAI	Bramantip	P is not empty
	EAO	Fesapo	M is not empty

conclude that "No Ss are P," in which case, we recognize the argument as an instance of Celarent, in the first figure. For this reason, forms such as *EAO* in the first figure, where the conclusion is weaker than what the premises support, are referred to as *weakened moods*. While technically valid, they are generally ignored.

If you care to check, there are four other weakened moods: *AAI* in the first figure, *AEO* and *EAO* in the second figure, and *AEO* in the fourth figure. They are unworthy even of receiving their own table.

3.5 Aristotle's Proof Methods

We have noted that the argument

> All mammals are animals.
>
> All cats are mammals.
> ———————————
> All cats are animals.

really seems very elementary. Aristotle agreed. He treated this as an instance of a *perfect syllogism*, by which he meant an argument form so obviously correct that anyone doubting it would be thought in some way perverse. In modern terminology, we would say Aristotle takes the correctness of the perfect syllogisms as axiomatic. He understood the perfect syllogisms to be the valid forms in the first figure.

In Aristotle's view, whereas the perfect syllogisms seem too obvious to deny, the imperfect syllogisms require proof. He employed two techniques in this regard.

The first is to apply the proposition conversions we studied in Section 3.3 to show that a given imperfect syllogism is actually equivalent to some perfect syllogism.

Let us see how this might play out for the second-figure forms. Here is an instance of Cesare:

> No cats are fish.
>
> All sharks are fish.
> ———————————
> No cats are sharks.

By converting the first premise, we obtain:

> No fish are cats.
>
> All sharks are fish.
> ———————————
> No sharks are cats.

which is an instance of the first-figure form Celarent. However, since our original argument had "cats" as the major term and "sharks" as the minor, Aristotle would insist on converting the conclusion to "No cats are sharks."

Cesare has an *E* statement for its major premise and an *A* statement for its minor premise. Camestres differs only by reversing these roles, so that the *A* statement is the major premise and the *E* statement the minor. Consequently, converting the *E* statement, followed by reversing the order of the premises, will once again produce an instance of Celarent.

Something new happens when we consider Festino. Given

> No cats are terriers.
>
> Some mammals are terriers.
> ———
> Some mammals are not cats.

we convert the major premise, thereby producing an instance of the first-figure form Ferio:

> No terriers are cats.
>
> Some mammals are terriers.
> ———
> Some mammals are not cats.

The remaining second-figure form is Baroco, but this time no simple conversion suffices to reduce it to a first-figure form. This leads us to the second of Aristotle's proof techniques, which we would today refer to as a proof by contradiction. The idea is to prove that a proposition is true by showing that its negation leads to a contradiction.

Let us apply this to an instance of Baroco:

> All terriers are dogs.
>
> Some animals with tails are not dogs.
> ———

As suggested by Table 3.4, the correct conclusion is "Some animals with tails are not terriers."

To see that this is correct, suppose instead that the conclusion is false. It would follow that "All animals with tails are terriers." Together with "All terriers are dogs," we now have premises corresponding to the perfect form *Barbara*, leading to this argument:

> All animals with tails are terriers.
>
> All terriers are dogs.
> ———
> All animals with tails are dogs.

This new conclusion directly contradicts the original premise, "Some animals with tails are not dogs." Since the proposed conclusion "All animals with tails are terriers" leads to a contradiction, we conclude that its negation, "Some animals with tails are not terriers," must be correct.

We have already seen that the sequence of the three vowels in each mnemonic name encodes the mood of the argument. We can now see there is considerably more to be gleaned from these names. The first-figure forms start with the letters B, C, D, and F. The first letter of each of the second-figure forms matches the first letter of the first-figure form to which it can be reduced. We have seen that Cesare and Camestres reduce to Celarent, that Festino reduces to Ferio, and that Baroco reduces to Barbara. Moreover, we find that an "s" in the mnemonic indicates that the reduction is achieved, at least in part, by a simple conversion of the "e" premise preceding the "s." This applies to Camestres and Festino, as we have seen. An "m" indicates that the order of the premises must be swapped to obtain a first-figure form. Finally, a "c" in the middle of the word indicates that a proof by contradiction will be used.

Using this information, we can see how to convert the third-figure forms to those in the first figure:

- Disamis expresses the mood *IAI*. It reduces to the first-figure form Darii by a simple conversion of the major premise, followed by swapping the order of the premises.
- Datisi expresses the mood *AII*. It reduces to Darii by a simple conversion of its minor premise.
- Bocardo expresses the mood *OAO*. We employ a proof by contradiction that leads to an instance of the first-figure form Barbara.
- Ferison expresses the mood *EIO*. It reduces to the first-figure form Ferio by a simple conversion of its minor premise.

It can be more enjoyable than you might think to work out concrete examples of these reductions, so have a go at it!

Just as interesting as Aristotle's methods for establishing the validity of syllogisms is a technique he used to show that certain pairs of premises fail to syllogize.

A modern approach for showing that an argument form is not valid is to construct an instance of an argument in that form with true premises and a false conclusion. This is essentially what Aristotle did, but he expressed himself in an interesting way:

> But if M is predicated of all O, but not of all N, there will be no syllogism. Take the terms animal, substance, raven; animal, white, raven. Nor will there be a conclusion when M is predicated of no O, but of some N. Terms to illustrate a positive relation between the extremes are animal, substance, unit; a negative relation, animal, substance, science. (Aristotle 2001, 71–72)

The term "substance" has a technical meaning in Aristotelian metaphysics, but for our purposes, we can think of it as equivalent to "thing."

Aristotle's example is meant to show that premises of the form:

All Ss are Ms.

Some Ps are not Ms.

do not lead to a syllogism. To see this, Aristotle asks us to consider these premises:

All ravens are animals.

Some substances are animals.

We know empirically (as opposed to logically) that the correct relationship between "ravens" and "substances" is "All ravens are substances." Now consider these premises:

All ravens are animals.

Some white things are animals.

This time the empirically correct conclusion is that "No ravens are white."

The arrangement of the middle term with respect to the extremes places these arguments in the second figure, but the first argument is in the mood *AOA*, while the second is in the mood *AOE*. Though the premises have the same logical form, in one case they lead to a universal affirmative conclusion, while in the other they lead to a universal negative. This shows that no general conclusion can be drawn from premises of this form. Clever!

Folks, there is more, oh so much more, that could be said about the nuances of Aristotelian logic, but I fear your patience may, at this point, be wearing thin. So, let us move on now to Carroll's work, and marvel at his ability to turn all this into a game.

3.6   Notes and Further Reading

Many books and online resources present the basics of syllogistic logic. A few that I found especially helpful in preparing this chapter are the books by Bird (1964), Rose (1968), Alexander (1969), and Flage (1994, Ch. 4), as well as the article by Robin Smith (2017) for the *Stanford Encyclopedia of Philosophy.* Of course, the full text of the *Organon* is freely available online and is well worth a look, despite the density of the prose.

The subtleties of Aristotle's logic have occupied philosophers and historians for quite a long time. Unsurprisingly, almost all facets of his writing are the subject of vigorous debate and discussion in the scholarly literature. If you wish to go more deeply into this subject, you could start with the book by Lear (1980) or the article by Leszl (2004).

CHAPTER 4

The Empuzzlement of Aristotelian Logic

We need a word for the process through which flaccid and boring text-book discussions of difficult material are transformed into fun and engaging puzzles. I suggest "empuzzlement" for that purpose. If ever there was a subject in need of empuzzlement, it is Aristotelian logic.

As I mentioned at the start of Chapter 3, Lewis Carroll (1832–1898) published a short book called *The Game of Logic* in 1886. In this work, he used Aristotelian logic to devise puzzles intended for young people. Where Aristotle saw a complete and systematic account of the entirety of human reasoning, Carroll saw a game suitable for children.

"Lewis Carroll" was a pseudonym used by Charles Dodgson, a British mathematician. As Dodgson, he was a respected scholar who made important contributions to linear algebra, geometry, and voting theory. As Carroll, however, he had a second career producing works intended for a broader audience. Most famously, he is the author of the children's stories *Alice's Adventures in Wonderland* and *Through the Looking-Glass*.

While Aristotle properly gets the credit for inaugurating the study of formal logic, it can be fairly said that Carroll created the field of recreational logic. Certainly, examples of logic puzzles existed prior to Carroll. However, it fell to Carroll to show that puzzles could be an important pedagogical tool for presentations of logic. His puzzles are meant not only to amuse, but also to educate.

4.1 Diagrams for Propositions

As is often the case with children's games, some setup is required before we can play.

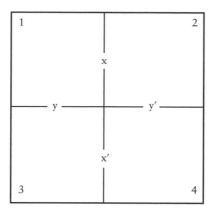

Figure 4.1. A Venn diagram based on rectangles instead of circles. The small numbers in the corners are not strictly part of the diagram. They are included only so that individual regions of the diagrams can be identified readily in the text discussion.

Central to Carroll's game was the use of diagrams to represent propositions. Rather than the traditional circles used by Venn, Carroll used diagrams based on rectangles. This is shown in Figure 4.1.

In the figure, we think of x and y as representing properties an object might have or sets to which they belong. The notations x' and y' then denote lacking that property or not belonging to that set.

For example, if we think of x as denoting cats and y as denoting fish, then region 1 contains all objects that are both cats and fish. This region is empty, of course. Region 2 represents all the cats that are not fish, and this is where all the cats reside. Region 3 represents all the fish that are not cats, which is to say, this region contains all the fish. Finally, region 4 contains those things that are neither cats nor fish.

A syllogism involves three terms, not two, so a larger figure is needed to represent it. This is shown in Figure 4.2. In addition to x and y, we now use m to denote the middle term. Objects possessing m reside inside the small square, while those lacking m are outside the small square. Thus, for example, if x and y are cats and fish as before, while m represents being orange, then those objects that are cats, fish, and orange reside in region 9. Region 8 contains everything that is neither cat, nor fish, nor orange, and region 10 represents orange cats that are not fish.

Carroll regarded his rectangular diagrams as an important improvement over Venn's circles, for two reasons.

The first was that rectangular figures are easily adapted to handle larger numbers of sets, whereas using circles or ellipses (as Venn did)

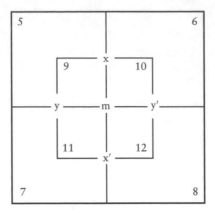

Figure 4.2. Carroll's diagram for representing the three terms of a syllogism. Again, the small numbers are included only to permit easy identification of different regions.

quickly leads to unmanageable diagrams. (This point will be explored in Section 5.1.)

The second was that with n terms, there are 2^n ways to have or to lack the various properties. With circles, however, only 2^{n-1} closed regions are possible. That is, two overlapping circles enclose three regions instead of the desired four, three overlapping circles can enclose at most seven regions instead of eight, and so on. The area outside the circles is left to represent those objects lacking all of the properties, but Carroll thought it a weakness of Venn diagrams that such objects did not have a closed region of their own. Modern textbooks typically enclose their Venn diagrams within a large rectangle, specifically to address this issue, but no doubt Carroll would have considered that inelegant. His rectangular diagrams had the full complement of regions, a fact that caused him considerable delight. In his 1896 book *Symbolic Logic*, he wrote:

> My Method of Diagrams *resembles* Mr. Venn's, in having separate *Compartments* assigned to the various Classes, and in marking these Compartments as *occupied* or *empty*; but it *differs* from his Method, in assigning a *closed* area to the *Universe of Discourse*, so that the Class which, under Mr. Venn's liberal sway, has been ranging at will through Infinite Space, is suddenly dismayed to find itself "cabin'd, cribb'd, confined," in a limited Cell like any other Class! (Carroll 1958, 176)

That is both an excellent example of Carroll's often lighthearted prose, as well as his obsessiveness over details.

4.2 Playing the Game

The "game" in *The Game of Logic* is as follows: The player requires two boards, one depicting the diagram for three terms, the other depicting the diagram for two. Carroll referred to these, respectively, as the *triliteral* and *biliteral* diagrams. The player also needs two different sets of counters, sized so that they can fit comfortably in the regions of the diagram. Carroll suggested using red and gray counters for this purpose, but since color printing is expensive, we shall use black and gray instead.

Carroll then presents the premises of a syllogism, typically with terms specifically chosen for their humorous effect. The object is to determine the deductive consequences of the premises. This is accomplished by representing the premises on the triliteral diagram, using black counters to indicate occupied regions and gray counters to indicate empty regions. Having extracted all the information latent in the premises, the player then determines what can be transferred from the triliteral to the biliteral, which is to say, that he seeks to eliminate the middle term. He then reads off the conclusion from the biliteral diagram. The game is now over, at least for this round.

Shall we play? Here is our first set of premises:

> No nice cakes are unwholesome.
>
> Some new cakes are unwholesome.
> _____
> Therefore, ??

Cakes, apparently, can have various combinations of niceness, newness, and wholesomeness. In our premises, wholesomeness is playing the role of the middle term. We seek a conclusion relating niceness to newness.

We begin with the triliteral diagram and define x, y, and m as follows:

$$x = \text{niceness},$$

$$y = \text{newness},$$

$$m = \text{wholesomeness}.$$

Our first premise is "No nice cakes are unwholesome." Now, nice cakes reside above the central, horizontal line, while unwholesome cakes are found outside the small square. Thus, if there are any nice, unwholesome cakes, then they reside in regions 5 and 6. Since the premise tells us that there are no such cakes, we place gray counters in each of those regions. The result is shown in Figure 4.3.

We now move to the second premise, "Some new cakes are unwholesome." As before, unwholesome cakes are found outside the small square.

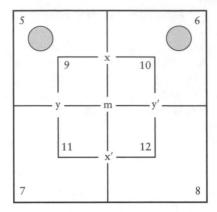

Figure 4.3. Recall that x represents niceness, y represents newness, and m represents wholesomeness. The gray counters therefore represent the premise "No nice cakes are unwholesome."

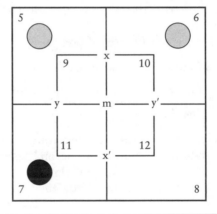

Figure 4.4. As before, x represents niceness, y represents newness, and m represents wholesomeness. Since the first premise implies that region 5 is empty, the new, unwholesome cakes described by the second premise must reside in region 7. Both premises are now represented on the same board.

New cakes are found to the left of the central vertical line. Cakes that are both new and unwholesome will therefore be found in regions 5 and 7. However, we already know that region 5 is empty. We conclude that the new, unwholesome cakes to which the premise refers are found in region 7, which therefore receives a black counter. Our board, which now represents both premises, is shown in Figure 4.4.

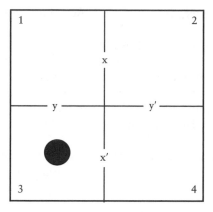

Figure 4.5. The biliteral diagram filled in with the information available from Figure 4.4.

Now that the information contained in the premises is represented in our triliteral diagram, we ask: What can be transferred to the biliteral diagram? This diagram contains four regions corresponding to combinations of newness and niceness, but does not contain regions corresponding to wholesomeness. Let us consider the possibilities.

Do we have sufficient information to place a counter in region 1, which corresponds to cakes that are both nice and new? Sadly, we do not. Since region 5 in the triliteral has a gray counter, we know that such nice, new cakes as exist would have to be wholesome as well. However, since region 9 in the triliteral has no counter, we do not know whether there are such cakes.

By nearly identical reasoning, we find we have insufficient information to fill in region 2, which corresponds to cakes that are nice but not new. We plainly have no information with which to place a counter in region 4, since regions 8 and 12 have no counters at all.

This leaves only the black counter in region 7. It tells us there are actual cakes that are new but not nice. We can therefore transfer this counter from region 7 to region 3. We have now accounted for all four regions in the biliteral diagram, and the result is shown in Figure 4.5.

We see that the correct conclusion is, "Some new cakes are not nice." Our first round of the game is complete.

Of course, we could also have arrived at this conclusion by noting that this is an argument in the third figure with premises of the form *EI*. Referring back to Table 3.5 reminds us that this is an instance of Ferison, whose correct conclusion is an *O* statement relating new cakes to nice cakes. Thus, the conclusion is "Some new cakes are not nice," precisely as before.

However, Carroll's approach is more fun.

There are two general principles for transferring information from the triliteral to the biliteral diagram. Every region in the biliteral corresponds to two regions in the triliteral. For example, regions 5 and 9 in the triliteral correspond to region 1 in the biliteral, regions 6 and 10 correspond to region 2, and so on. Here are the two principles:

- A black counter in either of the two regions in the triliteral justifies placing a black counter in the corresponding region in the biliteral.
- A gray counter in the biliteral can only be justified by gray counters in both regions in the triliteral.

4.3   A Closer Look at Placing Counters

The premises in our first round of Carroll's game involved an *E* statement and an *O* statement—a universal negative and a particular affirmative. Statements of type *E* are easy to represent on the triliteral diagram: Given the terms S and M, there will be two regions where items possessing both attributes may reside. The statement "No Ss are Ms" then implies that there are gray counters in both regions.

After placing the *E* counters, we found it was easy to place the *O* counters. In principle, there were two regions in which new, unwholesome cakes could reside. However, one of them was known, from the first premise, to be empty, and this told us where to place our black counter.

Without the *E* statement, placing the counters for the *O* statement would have been problematic. Statements of type *I* and type *O* have existential import, implying that premises of this type require a black counter to be placed somewhere. However, in general, there will be two regions that could receive this black counter, and without further information, we do not know which to choose.

For example, how could we have represented the premise "Some nice cakes are wholesome" taken by itself?

Recall once more that *x*, *y*, and *m* represent niceness, newness, and wholesomeness, respectively. Nice cakes reside in the upper half of the square, while wholesome cakes are found inside the small square. Apparently, then, either region 9 or region 10 ought to receive a black counter. But which one? The proposition tells us there are nice, wholesome cakes, but we do not know whether those cakes are new or not new. Perhaps there are some of each.

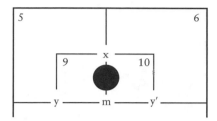

Figure 4.6. The black counter on the borderline means that at least one of the two sections is occupied, but we do not know which.

In this situation, Carroll suggests placing the counter right on the line between the two sections. This is shown in Figure 4.6 (which presents only the top part of the triliteral diagram). Carroll describes this approach as follows:*

> Our ingenious American cousins have invented a phrase to express the position of a man who wants to join one or other of two parties—such as their two parties 'Democrats' and 'Republicans'—but ca'n't make up his mind *which*. Such a man is said to be "sitting on the fence." Now that is exactly the position of the red counter you have just placed on the division-line. He likes the look of No. 9, and he likes the look of No. 10, and he doesn't know *which* to jump down into. So there he sits astride, silly fellow, dangling his legs, one on each side of the fence! (Carroll 1958, 9)

It remains to consider the representation of *A* statements—universal affirmatives. To make the discussion concrete, let us consider "All wholesome cakes are nice."

Plainly, this entails "No wholesome cakes are not nice." Therefore, we would place a gray counter in region 10. The disposition of region 9, however, is more difficult. We are once more forced to consider the question of existential import. Does "All wholesome cakes are nice" entail that wholesome cakes actually exist? Or does it only mean that wholesomeness cannot coexist with a lack of niceness? The former case would require a black counter in region 9, to accompany the gray counter in region 10. The latter would imply that the gray counter in region 10 represents the entire content of "All wholesome cakes are nice."

Carroll came down on the side of granting existential import to universal affirmative statements. He writes:

*The red marker that Carroll refers to in the following quote is shown in black in Figure 4.6.

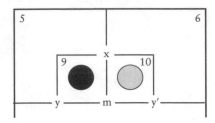

Figure 4.7. The proper representation of "All wholesome cakes are nice, according to Carroll."

[I]n every Proposition beginning with "some" or "all", the *actual existence* of the 'Subject' is asserted. If, for instance, I say "all misers are selfish," I mean that misers *actually exist*. If I wished to avoid making this assertion, and merely to state the *law* that miserliness necessarily involves selfishness, I should say "no misers are unselfish" which does not assert that any misers exist at all, but merely that, if any *did* exist, they *would* be selfish. (Carroll 1958, 19)

Therefore, the correct representation of "All wholesome cakes are nice" is the one shown in Figure 4.7.

As noted in Section 3.3, modern logicians are unanimous in denying existential import to *A* statements. Carroll's unwillingness to get with the program has been cited by some churlish modern reviewers of his work as evidence of his inadequacies as a logician (Alexander 1978). What matters for us, however, is that Carroll was clear about his assumptions.

As a further example, let us combine our principles by considering the premise "All nice cakes are not wholesome." Given Carroll's position on the existential import of universal affirmatives, we recognize this assertion as equivalent to two propositions:

- No nice cakes are wholesome.
- Some not nice cakes are not wholesome.

The first of these is represented by gray counters in regions 9 and 10 (since wholesome cakes are found inside the small square, while nice cakes are found above the central horizontal line). The second is represented by a strategically placed black counter. Since we lack the information to determine whether the counter belongs in region 5 or 6 (because we do not know whether our nice, unwholesome cakes are new or not new), we should place it on the border between the two regions. The final result is shown in Figure 4.8.

Let us close with one further observation about propositions, just to enjoy some more of Carroll's prose:

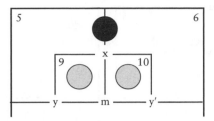

Figure 4.8. The proper representation of "All nice cakes are not wholesome, again, according to Carroll."

[W]hen a Proposition begins with "some" or "no," and contains more than two Attributes, these Attributes may be re-arranged, and shifted from one Term to the other, "*ad libitum*." For example, "some *abc* are *def*" may be rearranged as "some *bf* are *acde*," each being equivalent to "some Things are *abcdef*". Again "No wise old men are rash and reckless gamblers" may be re-arranged as "No rash old gamblers are wise and reckless," each being equivalent to "No men are wise old rash reckless gamblers." (Carroll 1958, 19)

4.4 One More Example

Let us proceed, more quickly this time, through one further example. Here are its premises:

> All wasps are unfriendly.
>
> All puppies are friendly.
> _____
> Therefore, ??

Since both premises are of type *A* (universal affirmatives), it behooves us, following Carroll's example, to write out explicitly what they entail. The first premise is equivalent to the pair of statements:

- No wasps are friendly.
- There exist wasps that are unfriendly.

The second premise is equivalent to

- No puppies are unfriendly.
- There exist puppies that are friendly.

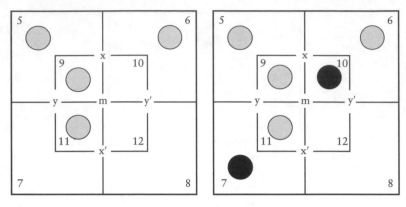

Figure 4.9. Recall that x represents being a puppy, y represents being a wasp, and m represents friendliness. On the left, the gray counters represent the propositions "No wasps are friendly" and "No puppies are unfriendly." On the right, the black counters represent the new information provided by "There exist wasps who are unfriendly" and "There exist puppies who are friendly."

We now define the following notation:

$$x = \text{being a puppy,}$$

$$y = \text{being a wasp,}$$

$$m = \text{friendliness.}$$

Since "friendliness" is the middle term, it made sense to represent it with m. However, our decisions regarding what x and y represent were arbitrary.

Now, we effectively have four premises with which to work, but since E statements (universal negatives) are simpler to represent on our triliteral diagram, let us start with them.

Friendly wasps, if any existed, would be found in regions 9 and 11. Since the premise implies that there are no friendly wasps, each of those regions receives a gray counter. Likewise, all unfriendly puppies would reside in regions 5 and 6, so these regions receive gray counters as well. The result is shown on the left in Figure 4.9.

We are also told that unfriendly wasps exist. Since such wasps can only be found in regions 7 and 11, and since region 11 is already known to be empty, we conclude that region 7 should receive a black counter. Meanwhile, friendly puppies are only found in regions 9 and 10. Since

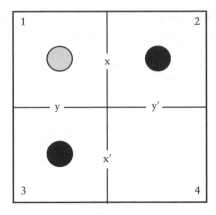

Figure 4.10. The small diagram that follows from the large one depicted on the right in Figure 4.9.

9 already has a gray counter, we see that region 10 must have a black counter.

Once these black counters have been added to our diagram, the result is as shown on the right in Figure 4.9.

This diagram is certainly busier than the one in our previous example! Carroll now instructs us to transfer what we can from the triliteral to the biliteral diagram. We have much to transfer: The gray counters in regions 5 and 9 imply we can place a gray counter in region 1, while the black counters in regions 7 and 10 imply there should be black counters in regions 3 and 2, respectively.

That is readily done, leading to the biliteral diagram shown in Figure 4.10.

From here it is easy to read off the conclusion "No puppies are wasps." Of course, we could as easily render the conclusion in the form "No wasps are puppies," if we prefer.

So ends round two.

4.5 Are We Having Fun Yet?

Carroll published *The Game of Logic* under his pen name, famous from its association with the *Alice* books, precisely to make clear his intention to popularize logic among children. The book opens with a short poem addressed "To my Child-Friend," and Carroll emphasizes the importance of seeing his work as a game.

In a brief preface, Carroll notes that an advantage of his game as contrasted with others is the possibility of playing it with just a single player, thereby making it simpler to find the necessary resources. (Carroll also notes that the game is more fun if several play at once, so that each may correct the errors of the others.) He then writes:

> A second advantage, possessed by the Game, is that, besides being an endless source of amusement (the number of arguments, that may be worked by it, being infinite), it will give the Players a little instruction as well. But is there any great harm in *that*, so long as you get plenty of amusement? (Carroll 1958, Preface)

In a preface to the first volume of his later work *Symbolic Logic*, the early parts of which reprise the material of *The Game of Logic*, Carroll writes, after discussing the manner in which a reader should work her way through the book:

> If, dear Reader, you will faithfully observe these Rules, and so give my little book a really *fair* trial, I promise you, most confidently, that you will find Symbolic Logic to be one of the most, if not *the* most, fascinating mental recreations! . . .I have myself taught most of its contents, *viva voce*, to *many* children, and have found them to take a real intelligent interest in the subject. (Carroll 1958, xvii)

I confess to being rather taken with the game myself. After reading the book, I made triliteral and biliteral diagrams and grabbed piles of pennies and nickels to use as counters. I then worked my way through a dozen or so of Carroll's puzzles. Once I had the hang of representing premises on the diagrams, I found it fascinating simply to watch the conclusion just appear, as if by magic, on the board.

However, it is an occupational hazard for mathematicians both to overestimate the pleasure their subject holds for other people and to underestimate the difficulties they find in appreciating our craft. An anonymous reviewer for the then-important publication *Literary World* was unimpressed with Carroll's work. I reproduce his review in its entirety. Let me preface this by noting Dr. Blimber is a character from Charles Dickens's novel *Dombey and Son*. You can infer what sort of person he is from the way he is referenced here.

> We confess to having spent some minutes in trying to make out just how children are to be persuaded to enjoy Mr. Lewis Carroll's new book, *The Game of Logic*, with its accompanying diagrams and red and gray wafers.

There *may* be young people capable of being amused by such syllogisms as "No old rabbits are greedy," "Some not greedy rabbits are black," "All white rabbits are free from greediness," and by disposing *x* and *y*, predicates, attributes, major and minor premises, in order due, with a red wafer here and two gray wafers there; but we should be at a loss where to lay our hands upon such young people, outside of the establishment of the late well-known Dr. Blimber. We seem to see some pale little Dombey junior bending a puzzled brow over the book, and trying to convince himself that it is fun and a game, and not hard work under a thin disguise; but a sturdy boy, not of the little Paul order and not educated by Dr. Blimber, would, we are inclined to think, spurn *The Game of Logic* as a stupid sham, black rabbits, greedy rabbits, pink pigs, and all, and clamor for some play that is really play, or else some study that is really study, on the principle that two things, each good in itself, often make when mixed a third thing which is neither good nor desirable. (Anonymous 1887, 122)

I have enough experience as a math teacher to know that my students often react with indifference to things I regard as beautiful, and not because of any lack of understanding on their part. The reviewer is certainly not obligated to like the things we math-types like, though his tone seems more snide than necessary.

That said, there are certain puzzles presented in *The Game of Logic*, and also in Carroll's later work *Symbolic Logic*, that really make you wonder about Carroll's notions of fun. Let me close this chapter by reproducing for you an actual puzzle that Carroll offers at the end of *The Game of Logic*. It made me think the anonymous reviewer might have had a point in wondering where to find children, or anyone else, who would think it fun to work this out. Carroll does not present a solution to this puzzle, and neither shall I. Here we go:

Extract from the following speech a series of Syllogisms, or arguments having the form of Syllogisms: and test their correctness.

It is supposed to be spoken by a fond mother, in answer to a friend's cautious suggestion that she is perhaps a little overdoing it, in the way of lessons, with her children.

Well, they've got their own way to make in the world. We can't leave them a fortune a piece! And money's not to be had, as *you* know, without money's worth: they must *work* if they want to live. And how are they to work, if they don't know anything? Take my word for it, there's no place for ignorance in *these* times! And all authorities agree that the time to learn is when you're young. One's got no memory afterwards, worth speaking of. A child will learn more in an hour than a grown man in five. So those, that have to learn, must learn when they are young, if ever they're to learn at all. Of course that doesn't do unless children are *healthy*: I quite allow *that*. Well, the doctor tells me no children are healthy unless they've got a good colour in their cheeks. And only just look at my darlings! Why, their cheeks bloom like peonies! Well, now, they tell me that, to keep children in health, you should never give them more than six hours altogether at lessons in the day, and at least two half-holidays in the week. And that's *exactly* our plan, I can assure you! We never go beyond six hours, and every Wednesday and Saturday, as ever is, not one syllable of lessons do they do after their one o'clock dinner! So how you can imagine I'm running any risk in the education of my precious pets is more than *I* can understand, I promise you! (Carroll 1958, 95–96)

While we can probably all agree it would be more tedious than fun to extract all possible categorical propositions and syllogisms from that passage, I *do* think it makes for enjoyable reading just for its own sake.

4.6 Puzzles for Solving

The Game of Logic contains dozens of humorous premises from which the reader is expected to draw conclusions. Our first puzzle contains three examples.

Puzzle 13. *For each of the following pairs of premises, deduce a conclusion that does not involve the middle term:*

  (a) *Some eggs are hard boiled; No eggs are uncrackable.*
  (b) *No monkeys are soldiers; All monkeys are mischievous.*
  (c) *All pigs are fat; No skeletons are fat.*

(Carroll 1958, 53)

Carroll also devised puzzles in which the reader has to extract the premises from a humorous story. I close this chapter by presenting two examples.

Puzzle 14. *Extract pairs of premises out of each of the following fragments, and deduce a conclusion if there is one:*

(a) "The Lion, as any one can tell you who has been chased by them as often as I have, is a very savage animal: and there are certain individuals among them, though I will not guarantee it as a general law, who do not drink coffee."

(b) "They say no doctors are metaphysical organists: and that lets me into a little fact about you, you know."

"Why, how do you make that out? You've never heard me play the organ."

"No, doctor, but I've heard you talk about Browning's poetry: and that showed me that you're metaphysical, at any rate. So—"

(Carroll 1958, 92–93)

4.7 Solutions

Puzzle 13a. Let m represent eggs, x represent hard boiled, and y represent crackable. The triliteral and biliteral diagrams for the premises are shown in Figure 4.11. From the biliteral diagram, we read off that the correct conclusion is "Some hard-boiled things can be cracked."

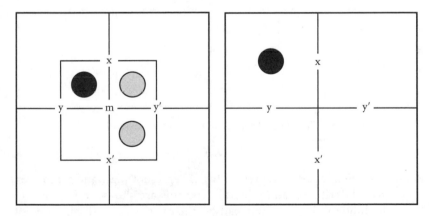

Figure 4.11. The triliteral and biliteral diagrams for Puzzle 13a.

Puzzle 13b. This time, let m represent monkeys, x represent soldiers, and y represent mischievous. The triliteral and biliteral diagrams are shown in Figure 4.12. From the biliteral diagram, we read off that the correct conclusion is "Some mischievous creatures are not soldiers."

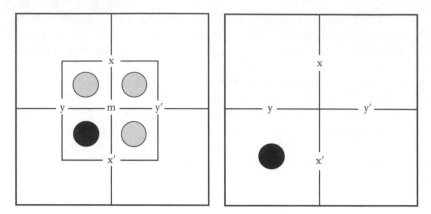

Figure 4.12. The triliteral and biliteral diagrams for Puzzle 13b.

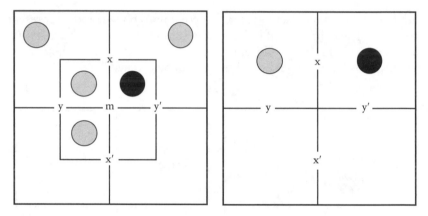

Figure 4.13. The triliteral and biliteral diagrams for Puzzle 13c.

Puzzle 13c. Finally, let *m* represent fat, *x* represent pigs, and *y* represent skeletons. The triliteral and biliteral diagrams are shown in Figure 4.13. From the biliteral diagram, we read off the conclusion "All pigs are not-skeletons." Given Carroll's views on existential import, this is equivalent to "Pigs exist, but none of them are skeletons."

Puzzle 14a. We can discern the premises "All lions are savage" and "Some lions do not drink coffee." From this, we can derive the conclusion "Some savage things do not drink coffee."

Puzzle 14b. This time we have the premises "No metaphysical doctors are organists" and "You are a metaphysical doctor." We conclude, "You are not an organist." If the reader objects that the second premise is not technically a categorical proposition in the precise sense we have defined (since it does not start with "All," "No," or "Some"), then keep in mind that we could have written, "All yous are metaphysical doctors." This is awkward phrasing, but it is logically equivalent to our original, more natural, premise.

CHAPTER 5

Sorites Puzzles

Carroll followed up *The Game of Logic* with a longer work titled *Symbolic Logic*, published in 1896. The contents of this work were heavily influenced by John Venn's characterization of the *elimination problem* in logic. Venn wrote:

> [T]he general problem ... may be stated thus:—'Given any number of propositions, of any kind, categorical, disjunctive, or otherwise, and involving any number of terms, find the mutual relation to one another, in respect of their extension, of any selection from amongst all these terms to any other such selection.' Again, the syllogism is a case of *Elimination* as well as of inference; that is, we know that we thus get rid, in our conclusion, of one term out of the three involved in our premises. (Venn 1881, xx–xxi)

Though Carroll's *Symbolic Logic*, like its predecessor *The Game of Logic*, contains many humorous puzzles, it was also meant to be a legitimate contribution to solving the elimination problem.

Rereading that last sentence, the phrase "contains many humorous puzzles" does not really do justice to the contents of *Symbolic Logic*. Carroll includes puzzles with up to 50 premises. Puzzles, mind you, that he seemed earnestly to believe his readers might enjoy solving. For myself, I am horrified by the thought even of transcribing such puzzles, much less trying to solve them. Thus, in this chapter, I confine my attention solely to those puzzles that illustrate Carroll's clever contributions toward solving the elimination problem.

Carroll intended for *Symbolic Logic* to be a three-volume work. Only the first volume was published in his lifetime. A nearly complete manuscript for the second volume was later discovered by philosopher

William Bartley, and this was published in 1977 (Bartley 1977). The topics we shall discuss are drawn from both parts.

5.1 A Quadriliteral Diagram?

In *The Game of Logic*, Lewis Carroll considered syllogisms, which are arguments with two categorical premises. An argument with more than two such premises is called a *sorites*, from a Greek word meaning "heap" or "pile." Presented with the premises to such an argument, how might we work out the conclusion?

Having spent so much time developing diagrams for representing syllogisms, you might wonder whether the technique can be extended to sorites puzzles. For example, what would you conclude from the following premises?

> No ducks waltz.
>
> No officers ever decline to waltz.
>
> All my poultry are ducks.

To convert these premises to symbolic form, we could define the following notation:

$$x = \text{ducks}, \quad y = \text{my poultry},$$
$$m = \text{officers}, \quad k = \text{willing to waltz}.$$

In *Symbolic Logic*, Carroll does not solve this (or any other) puzzle by using a quadriliteral diagram. However, in unpublished, handwritten notes, he does use such a diagram to provide a partial solution to this problem (Abeles 2010, 59; Moktefi 2013).

Let us take "creatures" to be our universe of discourse. With this notation, our premises can be expressed like this:

- No x are k.
- No m are not k.
- All y are x.

If we are to mimic the methods developed in *The Game of Logic*, then we need a diagram that can represent the possible relationships among four variables. This can be done as shown in Figure 5.1.

Notice that the diagram has 16 regions, corresponding to the 16 possible combinations of having or lacking the four attributes. The ease with which we accomplished this is another advantage of Carroll's rectangles

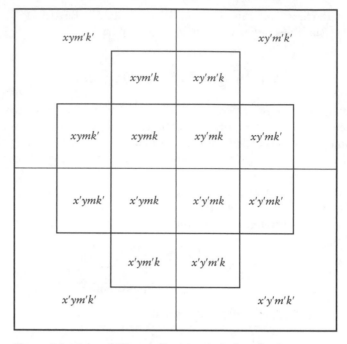

Figure 5.1. A quadriliteral diagram. As before, having or lacking x is represented, respectively, by being above or below the central horizontal line, while having or lacking y is determined in a similar manner by the central vertical line. Having or lacking m is represented, respectively, by being inside or outside the horizontal inner rectangle, while having or lacking k is represented in a similar manner by the vertical inner rectangle.

over Venn's circles. Venn was forced to use ellipses, instead of circles, to have the required number of regions (Venn 1881, 100–125).

Let us now represent our premises in the diagram. The first premise asserts that there is nothing that possesses both x and k. Thus, we need gray counters in the four regions above the central line and inside the vertical inner rectangle. The second premise corresponds to gray counters in any region possessing m, but lacking k. That is, we need gray counters in the four regions inside the horizontal inner rectangle but outside the vertical inner rectangle. The result is shown in Figure 5.2.

Recalling the discussion of A statements from Chapter 4, note that the third premise should be taken to entail two separate claims:

(1) Creatures that are my poultry exist.
(2) Nothing that is my poultry fails to be a duck.

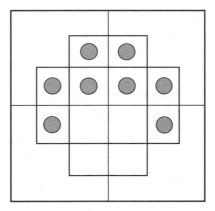

Figure 5.2. Representing the first two premises in the duck sorites.

Item (1) requires us to place a black counter in one of the eight regions corresponding to the possession of y (to the left of the central vertical line). However, item (2) tells us that nothing that possesses y fails to possess x. It follows that the black counter must be placed in a region corresponding to the possession of both x and y (above the central horizontal line and to the left of the central vertical line). Since three of the four possible regions already contain gray counters, the black counter belongs in the region corresponding to $xym'k'$. Moreover, there should be additional gray counters in any region corresponding to the possession of y but lacking x (to the left of the central vertical line but below the central horizontal line). The result is shown in Figure 5.3.

The conclusion to this sorites ought to relate y to m, since they are the terms that appear once each among the premises. The diagram includes gray counters in all regions corresponding to the possession of both y and m. Moreover, the one black counter indicates that creatures possessing y exist. It follows that the proper conclusion is "All y are not m," which translates to "All of my poultry are not officers."

The method worked! It was awfully cumbersome, however. It is no small task to keep track of what is represented by the diagram's various regions or to determine what follows from the riot of counters that we placed. And this is what happens when we deal with only three premises. Diagrams for more than three premises would quickly become so unwieldy that this method is not practical.

Carroll discusses the possibilities for diagrams capable of handling more than four terms. For example, to handle five terms, he suggests dividing each region of the quadriliteral diagram into an upper and lower half, to represent having or lacking the fifth attribute. This has the required 32 regions, but at the cost that the region for the fifth term is

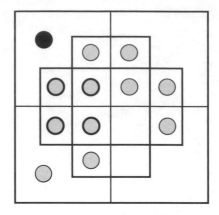

Figure 5.3. Representing all premises in the duck sorites. The correct conclusion is "All *y* are not *m*." The highlighted gray counters, along with the one black counter, are the ones relevant to drawing this conclusion.

not contiguous. Carroll goes all the way up to diagrams for ten terms, but he never used anything beyond the triliteral to solve an actual puzzle, at least not in those portions of his manuscript that have survived. Given the difficulties we experienced in using the quadriliteral diagram, it is unsurprising that the larger diagrams are of primarily theoretical interest.

Let us try to find a better method for solving these puzzles.

5.2 Notation and Formulas

Here is another of Carroll's puzzles:

(1) Every one who is sane can do logic.

(2) No lunatics are fit to serve on a jury.

(3) None of *your* sons can do logic.

Therefore, ??

Presumably, we should construe "is a lunatic" to mean "is not sane." The second premise can be rephrased to read, "No one who is not sane is fit to serve on a jury." With this understanding of the terms, the puzzle unravels quickly. The first and third premises produce an intermediate conclusion:

> Every one who is sane can do logic.
>
> None of *your* sons can do logic.
> _____
> None of *your* sons is sane.

It is convenient to rephrase this intermediate conclusion as "All of *your* sons are lunatics." We can now pair this with premise (2) to arrive at our final conclusion:

> All of *your* sons are lunatics.
>
> No lunatics are fit to serve on a jury.
> _____
> None of *your* sons is fit to serve on a jury.

We have solved the puzzle.

Let us refer to this approach as the "method of successive syllogisms": First, identify two premises from which we can draw an intermediate conclusion. Then pair this conclusion with an unused premise to draw a further conclusion. Continue in this manner until all premises have been used.

This method works in principle, but it might be difficult to apply when there are more than three premises, or when the premises contain more than two terms. It would be helpful to formalize the process so that it can be carried out mechanically. This is what Carroll accomplished, at least in part. His formalization involved two steps: the development of a compact notation for representing categorical propositions, and the derivation of formulas that could be appealed to for drawing intermediate conclusions.

Let us first consider Carroll's notation. If x is an attribute that may be possessed by some element of the universe, then Carroll would write x_1 to represent the proposition "Some existing things have the attribute x" (or more succinctly, "Some x exist"). Likewise, x_0 represents the proposition "No existing things have the attribute x," or simply, "No x exist."

If x and y are two attributes, then the propositions "Some things that are both x and y exist" and "Nothing that is both x and y exist" are denoted by xy_1 and xy_0, respectively. In each case, the subscript must be taken to apply back to the beginning of the expression.

To express "All x are y," recall our earlier discussion about existential import. In Carroll's usage, such a statement entails two distinct propositions: that some x exist, and also that all x are y. The assertion that all x are y is equivalent to the proposition that there are no x that are y'.

To express all of this in our subscript notation, we need a symbol for "and." Carroll used the dagger symbol (†) for that purpose. We can now express "All x are y" like this: $x_1 \dagger xy'_0$. This is cumbersome, however, and so we abbreviate it to this: $x_1y'_0$. Thus, we have the translation:

> $x_1y'_0$ means, "Things that are x exist, but
> things that are x but not y do not exist."

Note once more that each subscript is understood to apply back to the beginning of its expression. The subscript 1 applies only to x, while the subscript 0 applies to both x and y.

Carroll also employed the symbol $\mathbb{P}$ to indicate that a given set of premises entail a certain conclusion. Modern logicians typically use the turnstile symbol ($\models$) to express the same thing, but we shall follow Carroll's lead.

This might seem like a lot to absorb, but some examples will clarify matters. Given the premises "No x are m" and "All y are m," we conclude that "No x are y." In Carroll's notation, this becomes:

$$xm_0 \dagger y_1m'_0 \;\mathbb{P}\; xy_0.$$

Here is a second example: Given the premises "No x are m" and "Some y are m" we conclude that "Some y are x'." In Carroll's notation, this becomes:

$$xm_0 \dagger ym_1 \;\mathbb{P}\; x'y_1.$$

Carroll also employed some novel vocabulary. He referred to propositions that assert the existence of something, such as "Some x exist," as *Entities*. Propositions asserting that something does not exist were called *Nullities*. (Carroll always capitalized these words. Throughout this discussion, I have followed his conventions for capitalization.)

In a syllogism, the terms in the premises corresponding to Aristotle's middle term (which is to say the ones involving the attribute that does not appear in the conclusion), are the *Eliminands*. The terms appearing in the conclusion are the *Retinends*.

For example, in the argument:

> No x are m,
>
> All y are m,
> _____
> No x are y,

the Eliminand is m, while the Retinends are x and y. The first premise and the conclusion are Nullities, while the second premise is an Entity.

Table 5.1.
General formulas used by Carroll for solving sorites puzzles.

Figure I

$$xm_0 \dagger ym'_0 \; \mathbb{P} \; xy_0$$

(1) Two Nullities with Unlike Eliminands, yield a Nullity,
in which both Retinends keep their signs.

(2) A Retinend, asserted in the Premises to exist, may be so
asserted in the Conclusion.

Figure II

$$xm_0 \dagger ym_1 \; \mathbb{P} \; x'y_1$$

A Nullity and an Entity, with Like Eliminands,
yield an Entity, in which the Nullity-Retinend changes its Sign.

Figure III

$$xm_0 \dagger ym_0 \dagger m_1 \; \mathbb{P} \; x'y'_1$$

Two Nullities, with Like Eliminands asserted to exist,
yield an Entity, in which both Retinends change their Signs.

Finally, the presence or absence of the prime on a term, indicating things that either possess or do not possess that attribute, is referred to as the *Sign* of the term.

Carroll deduced four formulas for finding the correct conclusions to syllogisms. These formulas were used as intermediate steps when solving sorites puzzles. They are reproduced in Table 5.1, expressed using Carroll's notation and terminology. Following Aristotle, he describes his formulas as comprising different figures.

In the first row of Table 5.1, the Eliminands are said to be *Unlike*, because they occur with different signs. In the second row, the Eliminands are said to be *Like*, because they have the same sign, even though one appears as part of a Nullity while the other appears as part of an Entity. The Nullity-Retinend in this example is x. It is said to have *changed its Sign*, because in the Nullity premise it appears as x, but in the conclusion it appears as x'.

It appears to have been Carroll's intention that the reader would now stop to commit this table to memory, so that she could summon forth relevant formulas as needed when solving sorites puzzles. It will not reflect poorly on you, however, if you omit this step and instead proceed directly to the next section.

5.3 The Formalization in Action

Let us see how Carroll's formalization can be employed to solve our opening puzzle (see the start of Section 5.2).

Recall that the puzzle had these premises:

(1) Every one who is sane can do logic.
(2) No lunatics are fit to serve on a jury.
(3) None of *your* sons can do logic.

To apply Carroll's symbolism, first recast the premises in abstract form. To do this, define symbols for the various terms involved:

$$a = \text{able to do logic,}$$

$$b = \text{fit to serve on a jury,}$$

$$c = \text{sane,}$$

$$d = \text{your sons.}$$

Our universe of discourse can be taken to be "people." Let us use the notations a', b', c', and d' to denote, respectively, lacking a, b, c, or d. A statement like "Every a is b" can now be understood to mean, "All people who are able to do logic are fit to serve on a jury."

With this notation, the premises can be rendered as:

(1) All c are a.   (2) No c' is b.   (3) No d is a.

With the subscript notation, these premises can be further shortened to:

(1) $c_1 a'_0$,   (2) $c'b_0$,   (3) da_0.

Notice that the first and third premises are Nullities with Unlike Eliminands. After referring to Figure I of Table 5.1, we can write the intermediate syllogism as

$$c_1 a'_0 \dagger da_0 \; \P \; c_1 d_0.$$

Taking $c_1 d_0$ as a new premise, pairing it with our unused premise $c'b_0$, and again consulting Figure I of Table 5.1, leads to

$$c'b_0 \dagger c_1 d_0 \; \P \; db_0.$$

This translates back to "None of *your* sons is fit to serve on a jury," which we have already seen is the correct conclusion.

Let us try one further example, to ensure that you fully grasp the method. Following Carroll's lead, I encourage you to work this out for yourself before proceeding to the conclusion. Here are your premises:

(1) Puppies that will not lie still are always
   grateful for the loan of a skipping rope.
(2) A lame puppy would not say "Thank you"
   if you offered to lend it a skipping rope.
(3) None but lame puppies ever care to do worsted work.

As before, we begin by defining notation for the various attributes in the premises, taking the universe of discourse to be "puppies." The following seems reasonable:

$$a = \text{caring to do worsted work,}$$

$$b = \text{grateful for the loan of a skipping rope,}$$

$$c = \text{lame,} \quad d = \text{willing to lie still.}$$

If we construe "would not say 'Thank you' " to be equivalent to "not grateful," then our premises have the abstract form

(1) All d' are b.   (2) All c are b'.   (3) No c' are a.

In subscript notation, this becomes

(1) $d_1' b_0'$,   (2) $c_1 b_0$,   (3) $c' a_0$.

Our thorough command of Table 5.1 leads us to the following sequence of syllogisms:

$$d_1' b_0' \dagger c_1 b_0 \ \mathbb{P} \ d_1' c_0,$$
$$d_1' c_0 \dagger c' a_0 \ \mathbb{P} \ d_1' a_0.$$

Translated back to English, the conclusion is "All puppies who are willing to lie still do not care to do worsted work."

We worked out this conclusion by taking the premises in the order in which they were given, but we are not compelled to proceed in that manner. We could as easily have started with any premise, pairing it up with any other premise with which it forms a syllogism. For example, had we started with the second premise, we might have reasoned like this:

$$c_1 b_0 \dagger c' a_0 \ \mathbb{P} \ b a_0,$$
$$b a_0 \dagger d_1' b_0' \ \mathbb{P} \ d_1' a_0,$$

producing the same conclusion as before.

5.4 The Method of Underscoring

The method of successive syllogisms seems to get the job done, but it feels like we are still working too hard.

The argument given in Figure I in Table 5.1 is

$$xm_0 \dagger ym'_0 \ \mathbb{P} \ xy_0.$$

Notice that m is the Eliminand in this argument. To indicate this, let us rewrite the premises with the m terms underscored. Carroll recommends using a single underscore for the first occurrence of the Eliminand and a double underscore for the second. Here is the result:

$$x\underline{m}_0 \dagger y\underline{\underline{m}}'_0.$$

If we interpret the underscoring as indicating that those terms have been eliminated, then what survives are x and y, both occurring with the subscript 0. This technique allows for more compact solutions to sorites puzzles in which the premises all have the form given in Figure I.

To see how this works, let us return to the first example of Section 5.3:

(1) Every one who is sane can do Logic.
(2) No lunatics are fit to serve on a jury.
(3) None of *your* sons can do Logic.

Using the notation from Section 5.3, the premises can be written as follows:

$$(1) \ c_1a'_0, \quad (2) \ c'b_0, \quad (3) \ da_0.$$

Carroll notes that we can suppress the subscripts while working out the conclusion, for the following reason: The role of the subscripts is to tell us which terms, or combinations of terms, are Nullities and which are Entities. However, until we have a putative conclusion, we ignore the question of what exists and what does not. Instead, we care only about working out the proper signs for the Retinends. Once we have reached the conclusion, we will return to the premises to determine where to place the 0s and 1s.

Let us now write the premises on a line, separated by daggers and with the subscripts suppressed. Carroll recommends writing the numbers for the premises above the premises themselves. This helps keep us organized, especially in sorites puzzles with large numbers of premises. The result is:

$$\overset{1}{ca'} \dagger \overset{2}{c'b} \dagger \overset{3}{da}.$$

Taking the first two premises, we see that c can be taken as the Eliminand. Therefore, the c in the first premise receives an underscore, while the c in the second premise receives a double underscore. We then find that a can serve as the Eliminand in an argument with (1) and (3) as premises. So the a in the first premise receives an underscore, while the a in the third premise receives a double underscore:

$$\overset{1}{\underline{c}\,\underline{a}'} \,\dagger\, \overset{2}{\underline{\underline{c}}'b} \,\dagger\, \overset{3}{d\underline{\underline{a}}}$$

The surviving terms are b and d, neither of which was asserted to exist in the original premises. This leads to the conclusion bd_0, which is equivalent to db_0. Translated back into English this becomes "No son of *yours* is fit to serve on a jury," which is precisely the conclusion we reached before.

The second example from Section 5.3 was:

(1) Puppies that will not lie still are always
    grateful for the loan of a skipping rope.
(2) A lame puppy would not say "Thank you"
    if you offered to lend it a skipping rope.
(3) None but lame puppies ever care to do worsted work.

With the definitions given in Section 5.3, these premises became

$$(1)\ \ d_1'b_0', \quad (2)\ \ c_1b_0, \quad (3)\ \ c'a_0.$$

Once more applying the method of underscoring leads to

$$\overset{1}{d_1'\underline{b}_0'} \,\dagger\, \overset{2}{c_1\underline{\underline{b}}_0} \,\dagger\, \overset{3}{\underline{c}'a_0},$$

which entails the conclusion $d_1'a_0$, precisely as before.

To really appreciate the power of this method, however, let us see how it fares with a more complicated sorites puzzle. What conclusion follows from these premises?

(1) All the policemen on this beat sup with our cook.
(2) No man with long hair can fail to be a poet.
(3) Amos Judd has never been in prison.
(4) Our cook's cousins all love cold mutton.
(5) None but policemen on this beat are poets.
(6) None but her cousins ever sup with our cook.
(7) Men with short hair have all been in prison.

Translating these premises into subscript notation is no picnic, but Carroll helpfully shows the way. After declaring the universe of discourse to consist simply of "men," he suggests the following notation:

$a =$ Amos Judd,

$b =$ cousins of our cook,

$c =$ having been in prison,

$d =$ long-haired,

$e =$ loving cold mutton,

$h =$ poets,

$k =$ policemen on this beat,

$\ell =$ supping with our cook.

Translated into abstract form, our premises are now seen to be:

(1) All k are ℓ.   (2) No d are h'.   (3) All a are c'.

(4) All b are e.   (5) No k' are h.   (6) No b' are ℓ.

(7) All d' are c.

In subscript notation, we have

$$\overset{1}{k_1\ell_0'} \dagger \overset{2}{dh_0'} \dagger \overset{3}{a_1c_0} \dagger \overset{4}{b_1e_0'} \dagger \overset{5}{k'h_0} \dagger \overset{6}{b'\ell_0} \dagger \overset{7}{d_1'c_0'},$$

which becomes the following, when the subscripts are suppressed:

$$\overset{1}{k\ell'} \dagger \overset{2}{dh'} \dagger \overset{3}{ac} \dagger \overset{4}{be'} \dagger \overset{5}{k'h} \dagger \overset{6}{b'\ell} \dagger \overset{7}{d'c'}.$$

Let us now apply the method of underscoring. However, recall that this can only be done with premises whose Eliminands are of unlike sign. That means we will not be able to work with the premises in the order in which they were given.

Begin instead by seeking a premise that can be paired with the first. Specifically, we need a premise containing either k' or ℓ. The fifth premise is the first to answer to this description.

When the fifth premise is paired with the first, we are left with an intermediate conclusion containing ℓ' and h. Let us therefore continue our analysis by searching for a premise containing either ℓ or h'. The second premise will do nicely. By continuing in this manner, we obtain the following:

$$\overset{1}{\underline{k}\,\underline{\ell}'} \dagger \overset{5}{\underline{k}'\,\underline{h}} \dagger \overset{2}{\underline{d}\,\underline{h}'} \dagger \overset{6}{\underline{b}'\underline{\ell}} \dagger \overset{4}{\underline{b}e'} \dagger \overset{7}{\underline{d}'\,\underline{c}'} \dagger \overset{3}{a\underline{c}}$$

Thus a and e' are the Retinends, each appearing as part of a Nullity among the original premises. However, a was asserted to exist in the third premise. Consequently, the full conclusion is:

$$\overset{1}{k}\,\overset{}{\ell'} \dagger \overset{5}{\underline{k'}\,\underline{h}} \dagger \overset{2}{\underline{d}\,\underline{h'}} \dagger \overset{6}{\underline{b'}\underline{\ell}} \dagger \overset{4}{\underline{b}\underline{e'}} \dagger \overset{7}{\underline{d'}\,\underline{c'}} \dagger \overset{3}{a\underline{c}}\ \mathbb{P}\ e'a_0 \dagger a_1, \textit{that is, } \mathbb{P}\ a_1 e'_0.$$

Translated back into English, the conclusion is "Amos Judd loves cold mutton."

Not a bad trick!

However, in some sense, it really is just a trick. As already noted, the method of underscoring only works when the Eliminands are of unlike sign. Carroll tailored his examples to this limitation. Consequently, this method provides only a partial solution to the elimination problem, and nothing in Carroll's surviving works remedies this deficiency.

5.5 The Method of Trees

To this point, all of our sorites puzzles have involved biliteral premises, meaning that they only involved two terms. There is no reason to stop there, however. We could certainly consider puzzles with triliteral and multiliteral premises as well. Carroll's preferred method for solving such puzzles involved a combination of proof by contradiction and a clever sort of tree diagram for organizing his information. The idea is best understood by seeing how it applies in specific cases.

What can we conclude from the following premises:

$$\overset{1}{b'_1 a_0} \dagger \overset{2}{de'_0} \dagger \overset{3}{h_1 b_0} \dagger \overset{4}{ce_0} \dagger \overset{5}{d'_1 a'_0}\ ?$$

A quick inspection reveals that the Retinends will be c and h, as they are the terms that appear in only one premise each. The only issue, then, is to determine whether ch is a Nullity or an Entity.

To make that determination, Carroll started by assuming that ch was an Entity. The term ch now becomes the root of the simple tree diagram shown in Figure 5.4. This diagram was created by executing the following steps:

1. Look for the premise that involves c, which is premise 4. This premise asserts that c and e are incompatible. We therefore write e' underneath the ch, with a 4 next to it to remind us of which premise we used.
2. Next, look for the premise that involves h, which is number 3. This premise asserts that h and b are incompatible. We therefore write

$$\boxed{\begin{array}{c} ch \\ 3,4.\ e'b' \\ 1,2.\ d'a' \\ 5.\bigcirc \end{array}}$$

Figure 5.4. The tree that results when determining the conclusion to be drawn from the premises $b'_1 a_0 \dagger de'_0 \dagger h_1 b_0 \dagger ce_0 \dagger d'_1 a'_0$.

b' next to the e', and add a 3 next to it to remind us that we used premise 3 at this step.

3. Having extracted all of the immediate inferences from our assumption that ch is an Entity, let us now move on to the second line of the tree. Where do we find b' and e' among the premises? They are found in premises 1 and 2, which tell us that whatever possesses b' and e' must also possess a' and d'. These attributes are placed on the next line of the tree, numbered with their corresponding premises.

4. The final step is to note that premise 5 tells us that nothing possesses both d' and a'. Thus, our assumption that ch was an Entity leads to a contradiction. We indicate this by closing the tree with a large circle to indicate that the process has now terminated. Since the assumption that ch was an Entity has led to a contradiction, we know that the correct conclusion is that ch is a Nullity. Moreover, premise 3 tells us that h exists. Therefore, the full conclusion is $h_1 c_0$.

The process by which the tree was created might seem confusing, but it becomes easier to understand if we follow a suggestion from Carroll himself and imagine saying everything out loud:

> Let us start by assuming that ch is an Entity. Premises 3 and 4 tell us that anything that possesses both c and h must also possess b' and e'. Premises 1 and 2 tell us that anything that possesses both b' and e' also possess both a' and d'. We have now shown that if anything exists that is both c and h, then this object is also a' and d'. But premise 5 tells us that there is nothing that is both a' and d'. This is a contradiction, forcing us to conclude that ch is a Nullity. Since premise 3 ensures that objects possessing h exist, the full conclusion is $h_1 c_0$.

Our diagram for this puzzle did not look like a tree, since there was no branching. Let us consider a more difficult example involving triliteral propositions.

TABLE 5.2.
The register for the premises in Puzzle 15.

a	b	c	d	e	h	k	ℓ	m
7	4, 7	5		6	1, 3, 6	1, 5	4	1
3		2, 6	2	2	4	3, 7		5

Puzzle 15. *What would you conclude from these premises:*

$$\overset{1}{hm_1k_0} \dagger \overset{2}{d'e'c_0'} \dagger \overset{3}{hk'a_0'} \dagger \overset{4}{b\ell_1h_0'} \dagger \overset{5}{ck_1m_0'} \dagger \overset{6}{hc'e_0} \dagger \overset{7}{ba_1k_0'}?$$

A quick scan reveals that d' and ℓ appear once each and therefore are Retinends. Moreover, b appears only in positive form and is therefore a Retinend as well. Everything else is an Eliminand.

We could now dive right in and start constructing our tree, but Carroll recommends that we employ a helpful bookkeeping device first. To keep track of which premises contain which terms, he constructed the diagram shown in Table 5.2. Carroll referred to a diagram of this sort as a *register*. The top row in each column records the premises in which the term appears in positive form, while the second row records the premises in which the term appears in negated form. For example, a appears in positive form in premise 7, and in negated form in premise 3. The term b appears in positive form in premises 4 and 7, and it does not appear anywhere in negative form. And so on.

We begin our tree by supposing that $bd'\ell$ is an Entity. This term is placed atop our tree. Next, observe that premise 4 contains two of the Retinends. This permits an immediate conclusion, specifically, that anything that possesses b and l must also possess h. Therefore, we place h on the next line, indexed with a 4.

But now things get tricky. There is no other premise that contains two of the Retinends. That means we are deprived of any further immediate conclusions. For example, we find that b appears in premise 7, which asserts that b is never found in the presence of both a and k'. That tells us that our hypothetical Entity, which possesses h, either lacks a or possesses k, but we do not know which.

No matter how we proceed from this point, we will be forced to consider separate cases. In our tree diagram, this is represented by a branching. Since we are forced to divide the tree, Carroll suggests beginning with the lowest-numbered premise that is useful, which is number 1. This premise tells us that h is never found in the presence of both m and

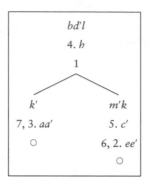

Figure 5.5. The tree that results when determining the conclusion to be drawn from the premises of Puzzle 15.

k. So, one branch of the tree will explore the consequences of assuming that our hypothetical Entity lacks k, while the other branch will consider the possibility that it possesses k but lacks m.

The completed tree is shown in Figure 5.5.

Carroll's tree method is a marvelous device for organizing the information to be gleaned from the premises of a sorites puzzle. It allows a solver to plow through a large mess of disorderly information in a short amount of time and in a small amount of space. Let us go one more round, to see how the method handles a large number of multiliteral premises.

Carroll works through the next example.

Puzzle 16. *What conclusion can be drawn from these premises:*

$$\overset{1}{an_1b_0'} \dagger \overset{2}{wm_1\ell_0'} \dagger \overset{3}{csn_0} \dagger \overset{4}{ar_1v_0'} \dagger \overset{5}{e_1c'\ell_0'} \dagger \overset{6}{mb_1t_0'} \dagger \overset{7}{k_1n_0'}$$

$$\overset{8}{dr_1a'e_0'} \dagger \overset{9}{rt_1w_0'} \dagger \overset{10}{e\ell_1'n_0'} \dagger \overset{11}{a's_0'} \dagger \overset{12}{db_1m_0'} \dagger \overset{13}{v_1e'k_0'} \dagger \overset{14}{bw_1'h_0'}?$$

The register for these premises is shown in Table 5.3.

A glance at the register reveals that d and r appear only in their positive form, while ℓ appears only in its negative form. Therefore, these are the Retinends, while all the other terms are Eliminands.

So, we hypothesize that $d\ell'r$ is an Entity and place it at the top of the tree. A survey of the register quickly reveals that we must divide right at the start, because there is no other premise that contains two Retinends and one additional term. The premises do not allow an immediate inference from the assumption that $d\ell'r$ is an Entity.

However, premise 8 contains two of the Retinends. It tells us that anything that is both d and r must also either be a, or it must lack a

TABLE 5.3.
The register for the premises in Puzzle 16.

a	b	c	d	e	h	k	ℓ
1, 4	12, 14	3	8, 12	5, 10	6	7	
8, 11	1	5		8, 13	14	13	2, 5, 10

m	n	r	s	t	v	w	
2, 6	1, 3	4, 8, 9	3	9	13	2	
12	7, 10		11	6	4	9, 14	

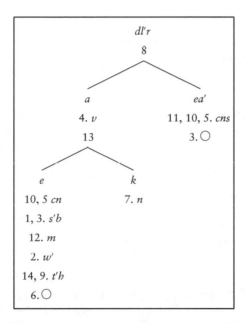

Figure 5.6. The tree corresponding to the premises of Puzzle 16.

and possess *e* (that is, it must be *ea′*.) We thus create a branching to explore separately the consequences of each assumption. Continuing in the same manner as in Puzzle 15 eventually leads to the tree shown in Figure 5.6.

Notice that there is a branch that does not terminate in a circle. Carroll notes that *n* also appears in the far-left branch of the tree, and that branch does terminate in a circle. Moreover, the manner in which that branch terminated made no later use of either the c next to the n, or the e just above it. Consequently, the termination of the far-left branch depends only on the presence of the n. We conclude that the same reasoning would also have shown the impossibility of the open branch; there is no need to reinvent the wheel. Consequently, $d\ell'r$ must be a Nullity. Since dr is an Entity (according to premise 8), the full conclusion is $dr_1\ell'_0$.

Having worked through three examples, we could certainly stop, having seen that the method of trees is an admirable device for organizing large amounts of information. We do not have to belabor this topic indefinitely.

However, I do not wish to conclude this section without communicating to you the enthusiasm with which Carroll pursued this topic. So, I will reproduce for you one final example, which he discusses in full in *Symbolic Logic*. He encourages his readers to go through everything in great detail, and you are certainly welcome to do so if you choose, but that is not my intention in reproducing this puzzle for you. Instead, I want you to enjoy it, and the tree to which it leads, simply as a painting. I think it is beautiful just as a thing to look at, even if you have no desire to work through the details.

Puzzle 17. *What can you conclude from these 24 premises?*

$$(1-5) \quad C\ell_1 E'_0 \; \dagger \; Av'_1 D_0 \; \dagger \; k_1 m'_0 \; \dagger \; \ell C'_1 (b'n')'_0 \; \dagger \; dsb_1 t'_0$$

$$(6-10) \quad tD_1 w'_0 \; \dagger \; dr'a'_1 A'_0 \; \dagger \; vw_1 B_0 \; \dagger \; em'_1 (r'b')'_0 \; \dagger \; Ha_1 c'_0$$

$$(11-15) \quad dtmav'_0 \; \dagger \; dst_1 A'_0 \; \dagger \; Dn'r'b'_1 z_0 \; \dagger \; cE'z_0 \; \dagger \; bs'_1 \ell' e'_0$$

$$(16-20) \quad atE_1 v'_0 \; \dagger \; rDh'_1 e'_0 \; \dagger \; mt'_1 D_0 \; \dagger \; An\ell'_1 c'_0 \; \dagger \; rdk'_1 h_0$$

$$(21-24) \quad ztB'_1 d_0 \; \dagger \; n\ell'_1 H'_0 \; \dagger \; Et'_1 z_0 \; \dagger \; dzrA'_1 a'_0$$

The register corresponding to these premises is given in Table 5.4.

The terms d, z, and D are plainly the Retinends, implying that dzD should be placed atop our tree. After much hard labor, the result is the tree shown in Figure 5.7.

The conclusion works out to be dzD_0, in case you were wondering.

Carroll presents a full solution to Puzzle 17. A hallmark of Carroll's writing is that, no matter how dry and technical the material might be, he always strove to make things as engaging as possible. To that end, he presented the solution to his tree problems by including a monologue that represented his precise thoughts as he worked through the premises. To give you a taste of Carroll's writing, here is a small portion of his

TABLE 5.4.
The register corresponding to the premises in Puzzle 17.

a	b	c	d	e	h
$10, 11, 16$	$4, 5, 9, 15$	14	$5, 7, 11, 12, 20, 21, 24$	9	20
$7, 24$	13	$10, 19$		$15, 17$	17

k	ℓ	m	n	r	s
3	$1, 4$	$11, 18$	$4, 19, 22$	$9, 17, 20, 24$	$5, 12$
20	$15, 19, 22$	$3, 9$	13	$7, 13$	15

t	v	w	z	A	B
$6, 11, 12, 16, 21$	8	8	$13, 14, 21, 23, 24$	$2, 19$	8
$5, 18, 23$	$2, 11, 16$	6		$7, 12, 24$	21

C	D	E	H
1	$2, 6, 13, 17, 18$	$16, 23$	10
4		$1, 14$	22

monologue for Puzzle 17. In this excerpt, he has just placed dzD at the top of the tree and is deciding how to proceed:

Now, what can we do with d? It occurs in $5, 7, 11, 12, 20, 21, 24$. Alas, they *all* divide! And so do the z's: and so do the great D's. Well, there's no help for it: we must divide at the very first start! Let's get a *biliteral* division, if we *can*. No. 21 is the first I can find, as it contains *two* Retinends: so it merely divides for t and B'.

[*I make a wide Branching under d: under the middle of the horizontal line I write 21, and under the two ends I write t' and B.*]

Now, is there any use tacking on t or B'? Let's see. Yes, t *can* be of further use, but B' of none.

[*I tack on t to B.*]

Now for the t'-Branch. 5 divides, but 18 doesn't: it gives us m'. And 23 gives us E'. That's a good beginning.

[*I write m'E' under t', with 23, 18, on the left.*]

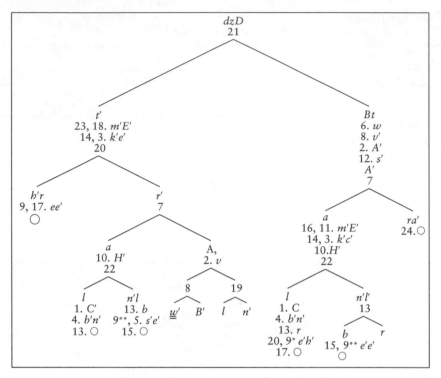

Figure 5.7. The tree diagram corresponding to the premises in Puzzle 17.

Now for the *Bt*-Branch. *B* only occurs in 8; and *that* divides. However, *t* helps us in 6, and gives us *w*: in all other Premises it divides.

[*I write w under Bt, with 6 on the left.*]

(Bartley 1977)

This goes on for several more pages!

5.6 *Puzzles for Solving*

Here are some more of Carroll's puzzles, so that you can try his methods for yourself. They make for amusing reading, even if you choose not to try to solve them. In each case, your task is to derive the strongest conclusion that follows from the premises. I have chosen five puzzles for this section, though Carroll provides no fewer than 60, all with full solutions, in his book.

Puzzle 18.

1. *Babies are illogical.*
2. *No person is despised who can manage a crocodile.*
3. *Illogical persons are despised.*

(Bartley 1977, 160)

Puzzle 19.

1. *Nobody who really appreciates Beethoven fails to keep silent while the Moonlight Sonata is being played.*
2. *Guinea pigs are hopelessly ignorant of music.*
3. *No one who is hopelessly ignorant of music ever keeps silent when the Moonlight Sonata is being played.*

(Bartley 1977, 165)

Puzzle 20.

1. *Things sold in the street are of no great value.*
2. *Nothing but rubbish can be had for a song.*
3. *Eggs of the Great Auk are very valuable.*
4. *It is only what is sold in the street that is really rubbish.*

(Bartley 1977, 166)

Puzzle 21.

1. *No shark ever doubts that it is well fitted out.*
2. *A fish that cannot dance a minuet is contemptible.*
3. *No fish is quite certain whether it is well fitted out, unless it has three rows of teeth.*
4. *All fishes, except sharks, are kind to children.*
5. *No heavy fish can dance a minuet.*
6. *A fish with three rows of teeth is not to be despised.*

(Bartley 1977, 173)

Puzzle 22.

1. *The only animals in this house are cats.*
2. *Every animal is suitable for a pet, that loves to gaze at the moon.*
3. *When I detest an animal, I avoid it.*
4. *No animals are carnivorous, unless they prowl at night.*
5. *No cat fails to kill mice.*
6. *No animals ever take to me, except what are in the house.*
7. *Kangaroos are not suitable for pets.*
8. *None but carnivora kill mice.*

9. *I detest animals that do not take to me.*

10. *Animals that prowl at night always love to gaze at the moon.*

(Bartley 1977, 175–176)

5.7 Notes and Further Reading

For the sake of completeness, let me mention that Carroll's method for tree construction included one further step that I have omitted from my discussion. After the tree was finished, Carroll carried out a check of its correctness by working his way through each branch, starting from the bottom and working up to the root. This involved adding some further notation to the tree. I do not think that this step is necessary for understanding Carroll's method, and I felt that including it would have belabored the proceedings beyond decency.

The papers by Francine Abeles (2005, 2007) and Amirouche Moktefi (2008) are helpful for understanding both the minutiae of Carroll's logic and the historical context in which he produced it. For a general survey of Carroll's logical and mathematical work, see Wilson and Moktefi (2019).

5.8 Solutions

The solutions that follow are precisely the ones Carroll provides in his book. For reasons that are unclear to me, Carroll often chose to omit certain letters as names of premises, but I have followed his notation nevertheless. To conserve space, I have not numbered the premises.

Puzzle 18. For the universe of discourse, use "persons." Now define the notation: $a=$ able to manage a crocodile, $b=$ babies, $c=$ despised, and $d=$ logical. The premises can now be expressed as

$$b_1 d_0 \dagger a c_0 \dagger d_1' c_0'.$$

After suppressing the subscripts and applying the method of underscoring, we get:

$$b\underline{d} \dagger \underline{d}' \underline{c}' \dagger a\underline{c} \, \mathbb{P} \, ba_0 \dagger b_1, \text{ that is, } \mathbb{P} \, b_1 a_0.$$

The conclusion is "Babies cannot manage crocodiles."

Puzzle 19. The universe is "creatures." Define the notation: $a=$ guinea pigs; $b=$ hopelessly ignorant of music, $c=$ keeping silent while the Moonlight Sonata is being played, and $d=$ really appreciating Beethoven. The premises can now be expressed as

$$dc'_0 \dagger a_1 b'_0 \dagger bc_0.$$

After suppressing the subscripts and applying the method of underscoring, we get:

$$d\underline{c}' \dagger \underline{b}\,\underline{c} \dagger a\underline{b}' \; \mathbb{P} \; da_0 \dagger a_1, \text{ that is, } \mathbb{P} \; a_1 d_0.$$

The conclusion is "Guinea pigs never really appreciate Beethoven."

Puzzle 20. The universe is "things." Define the notation: $a =$ able to be had for a song, $b =$ eggs of the Great Auk, $c =$ rubbish, $d =$ sold in the street, and $e =$ very valuable. The premises can now be expressed as

$$d_1 e_0 \dagger c' a_0 \dagger b_1 e'_0 \dagger c_1 d'_0.$$

After suppressing the subscripts and applying the method of underscoring, we get:

$$\underline{d}\,\underline{e} \dagger b\underline{e}' \dagger \underline{c}\,\underline{d}' \dagger \underline{c}'a \; \mathbb{P} \; ba_0 \dagger b_1, \text{ that is, } \mathbb{P} \; b_1 a_0.$$

The conclusion is, "An egg of the Great Auk is not to be had for a song."

Puzzle 21. The universe is "fishes." Define the notation: $a =$ able to dance a minuet, $b =$ certain that he is well fitted out, $c =$ contemptible, $d =$ having three rows of teeth, $e =$ heavy, $h =$ kind to children, and $k =$ sharks. The premises can now be expressed as

$$kb'_0 \dagger a'_1 c'_0 \dagger d'b_0 \dagger k'_1 h'_0 \dagger ea_0 \dagger d_1 c_0.$$

After suppressing the subscripts and applying the method of underscoring, we get:

$$\underline{k}\,\underline{b}' \dagger \underline{d}'\underline{b} \dagger \underline{k}'h' \dagger \underline{d}\,\underline{c} \dagger a'\underline{c}' \dagger e\underline{a} \; \mathbb{P} \; h'e_0.$$

The conclusion is, "No heavy fish is unkind to children."

Puzzle 22. The universe is "animals." Define the following notation:

$a =$ avoided by me; $b =$ carnivora; $c =$ cats;

$d =$ detested by me; $e =$ in this house; $h =$ kangaroos;

$k =$ killing mice; $\ell =$ loving to gaze at the moon;

$m =$ prowling at night; $n =$ suitable for pets; $r =$ talking to me.

The premises can now be expressed as

$$e_1 c_0' \dagger \ell_1 n_0' \dagger d_1 a_0' \dagger m' b_0 \dagger c k_0' \dagger c' r_0 \dagger h_1 n_0 \dagger b' k_0 \dagger r_1' d_0' \dagger m_1 \ell_0'.$$

After suppressing the subscripts and applying the method of underscoring, we get:

$$\underline{e}\,\underline{c}' \dagger \underline{c}\,\underline{k}' \dagger \underline{e}'\underline{r} \dagger \underline{b}'\underline{k} \dagger \underline{m}'\underline{b} \dagger \underline{r}'\underline{d}' \dagger \underline{d}a' \dagger \underline{m}\,\underline{\ell}' \dagger \underline{\ell}\,\underline{n}' \dagger h\underline{n}.$$

This entails $a'h_0 \dagger h_1$, which is equivalent to $h_1 a_0'$. The conclusion is, "I always avoid a kangaroo."

CHAPTER 6

Carroll's Contributions to *Mind*

Carroll made two further contributions to logic, in the form of two short papers published in the academic journal *Mind*. This is the sort of journal where scholars publish work addressed to one another, typically assuming considerable background knowledge in their readers. It is telling that even here Carroll published his logical work under his pseudonym, best known from the *Alice* books, which were directed at children. Even as he wanted to make a mark among professional logicians, he also strove for his work to be engaging and accessible.

In a discussion of Carroll's work on logic as a whole, Bertrand Russell remarked:

> I think he was very good at inventing puzzles in pure logic. When he was quite an old man, he invented two puzzles he published in a learned periodical, *Mind*, to which he didn't provide answers. And the providing of answers was a job, at least so I found it. (Moktefi 2008, 488)

Sometimes it is not easy to distinguish a fun and engaging puzzle from heavier fare meant to illuminate difficult issues. Even as Carroll presented his contributions to *Mind* as though they were the former, it quickly becomes clear they are also instances of the latter.

Let us have a look.

6.1 The Barbershop Puzzle

The first of Carroll's papers was called "A Logical Paradox" (Carroll 1894). The alleged paradox is presented in the form of a short dialogue between a teenager and two of his uncles. As with so much of Carroll's writing, it makes for whimsical reading regardless of what you think of its scholarly merits.

Since Carroll presented his puzzle in the form of a story, it is not possible to quote a small portion that captures all of the necessary details. The version of the puzzle that I will discuss is based on the lucid summary presented by Arthur Burks and Irving Copi (1950). However, I need to set up a few things before getting down to business.

Up to this point, we have been discussing Aristotelian logic, in which terms (or sets, or classes) enter into logical relationships with one another. In Part III, our focus will shift to propositional logic, in which propositions enter into such relationships. Carroll's two puzzles for *Mind* are based on questions about the correct interpretation of propositions employing the construction "if–then." More specifically, if p and q are propositions, then how is the truth value of the statement "if p, then q" related to the truth values of p and q individually?

As we shall discuss in Section 7.1, that is not an easy question. However, two things are clear. The first is that if I assume that the propositions "if p, then q" and "p" are both true, then I can conclude that q is true as well. This argument form is known as *modus ponens*, shortened from *modus ponendo ponens*, which translates roughly as "mode that affirms by affirming."

Likewise, if I assume that the propositions "if p, then q" and "not q" are both true, then I can assume that "not p" is also true. This is known as *modus tollens*, shortened from *modus tollendo tollens*, which translates roughly as "mode that denies by denying."

Carroll's story describes a barbershop staffed by three people named Allen, Brown, and Carr. At all times during business hours, at least one of them is required to be in the shop. From this, it follows that if Allen and Carr are out of the shop, then Brown must be in. However, it turns out that Allen prefers never to be alone, having recently been ill and being fearful of a relapse. As a result, any time Allen goes out, he takes Brown with him.

We can think of these conditions as representing rules that govern the movements of the three men. They can be represented like this:

(1) If Carr and Allen are out, then Brown is in.
(2) If Allen is out, then Brown is out.

What follows from these rules?

We might first note that (1) is logically equivalent to

(3) If Carr is out, then (if Allen is out, then Brown is in).

It would seem that the statements "If Allen is out, then Brown is out" and "If Allen is out, then Brown is in" are contraries. They cannot both be true. But since the first of these statements is given to be true as (2), we conclude that

(4) It is not the case that: If Allen is out, then Brown is in.

But now we apply modus tollens to statements (3) and (4), and discover that

(5) Carr is not out.

We seem to have shown that rules (1) and (2) imply that Carr can never be out. This is plainly absurd, because as long as either Brown or Allen is in, Carr can go out without any difficulty.

Very clever! How can we resolve this?

Shortly after the publication of Carroll's article, a solution was published by Warren Johnson. His solution opened like this:

> I presume Mr. Lewis Carroll's position to be that the problem raised by him in the last number of *Mind* presents a conflict between common sense and the rules of logic. It appears to me certain that the rules of logic when properly applied agree with the results of common sense. (Johnson 1894, 583)

In referring to the rules of logic, Johnson meant the manner in which logicians typically handle if–then statements. This approach is based on the idea that the statement "if p, then q" is really just saying that it is impossible for p to be true and q to be false simultaneously. Seen in this way, "if p, then q" is equivalent to "p is false or q is true." It follows that "if p, then q" is false only when p is true and q is false. It is true in every other case. This approach is known as "material implication."

There is a rich discussion to be had regarding the appropriateness of this approach, but it will be convenient to defer that discussion to Section 7.1. The point for now is that if we construe if–then statements in this way, then Carroll's paradox dissolves quickly.

Given our model for handling if–then statements, it is simply erroneous to assert that "If Allen is out, then Brown is out" and "If Allen is out, then Brown is in" are contraries. They *can* both be true, just so long as Allen is in. More generally, if the same antecedent is seen to imply contradictory conclusions, then that can be taken to imply that the antecedent is false.

Returning now to statement (3), we see that starting from the assumption that Carr is out, we are forced to conclude that Allen is in. It follows that "Carr and Allen are both out" is a contradiction. Therefore, our conclusion is "If Carr is out, then Allen is in, and if Allen is out, then Carr is in." This conclusion is quickly seen to be consistent with the given information.

Johnson's solution was quickly accepted as canonical. Bertrand Russell later endorsed it in his book *The Principles of Mathematics* (Russell 1903, 18).

This solution is certainly correct given the exigencies of material implication. However, we might wonder whether Johnson was correct to claim that this understanding of if–then statements comports with common sense. After all, if one person says "if p, then q," and another person says "if p then (not q)," then it sure *seems* like they are at odds with each other. That both can be right when p is false seems neither here nor there.

This observation forms the basis of the aforementioned paper by Burks and Copi:

> [I]f one politician argues that if the Conservatives win the election in 1950 then Britain's economic situation will improve, and another argues that if the Conservatives win in 1950 then Britain's economic situation will not improve, there is genuine disagreement. It would indeed be an over-zealous proponent of material implication who would expect the disputants to agree that another Labour victory at the polls would make both their statements true! (Burks and Copi 1950, 220)

They suggest a different solution to Carroll's puzzle. The conditional statements given in Carroll's premises, they argue, should be seen as instances of causal implication and not material implication. Whereas material implication holds that the truth of "if p, then q" is determined entirely by the truth values assigned to p and q treated as separate propositions, causal implication requires a causal connection between p and q. As an example of such an implication, Burks and Copi suggest the statement "If a is released, then it will fall," where a represents an object that is heavier than air and suspended over open space.

The statements "A causes B" and "A causes (not B)" are, indeed, contraries. They cannot both be true. That means that the Johnson/Russell solution to Carroll's puzzle does not work if we take the premises to be expressing causal implication as opposed to material implication. We might then wonder whether, given this interpretation, Carr really must remain inside after all.

This turns out not to be the case. Burks and Copi note that the argument to this conclusion is no more valid under causal implication than it is under material implication, but that the error lies in a different place. This time the problem comes at the very first step. Under material implication, the statements given above as (1) and (3) are logically equivalent. Under causal implication, they are not.

Burks and Copi provide a formal derivation of this conclusion, the details of which are tedious. The basic idea, however, can be seen by supposing we know that p and q together cause r. Would that imply that p by itself tells us anything about a causal relationship between q and r? The answer certainly seems to be no, and Burks and Copi confirm that intuition.

Carroll's barbershop puzzle was the subject of considerable discussion at the time, but it has no relevance for research today. As soon as we give a clear account for how we wish to interpret conditional statements, Carroll's premises can readily be analyzed in accord with that account. That said, Carroll's puzzle certainly does illustrate just how difficult it can be to come to terms with if–then statements.

6.2 Achilles and the Tortoise

Carroll's second contribution was called "What the Tortoise Said to Achilles" (Carroll 1895) and was written in the form of a lighthearted dialogue between the title characters.

After a brief introduction, Carroll presents three propositions:

(A) Things that are equal to the same are equal to each other.
(B) The two sides of this triangle are things that are equal to the same.
(Z) The two sides of this triangle are equal to each other.

In the story, the Tortoise instructs Achilles to write these statements down in his notebook. The following dialogue ensues, with the Tortoise speaking first:

"Readers of Euclid will grant, I suppose, that Z follows logically from A and B, so that any one who accepts A and B as true, *must* accept Z as true?"

"Undoubtedly! The youngest child in a High School ...will grant *that*."

"And if some reader had *not* yet accepted A and B as true, he might still accept the *sequence* as a *valid* one, I suppose?"

"No doubt such a reader might exist. He might say, 'I accept as true the Hypothetical Proposition that, *if* A and B be true, Z must be true; but I *don't* accept A and B as true.'..."

"And might there not *also* be some reader who would say 'I accept A and B as true, but I *don't* accept the Hypothetical'?"

"Certainly there might. ..."

"And *neither* of these readers," the Tortoise continued, "is *as yet* under any logical necessity to accept A as true?"

"Quite so," Achilles assented.

"Well, now, I want you to consider *me* as a reader of the *second* kind, and force me, logically, to accept Z as true." (Carroll 1895, 278–279)

Achilles warms to the challenge. He notes that the problem seems to be that the Tortoise is unable to make the inference from A and B to Z, because he does not accept the following conditional:

(C) If A and B are true, Z must be true.

The Tortoise suggests that C be written down in Achilles's notebook, so that the argument now proceeds from premises A, B, and C to the conclusion Z. We pick up the conversation after this point, with Achilles speaking first.

"If you accept A and B and C, you *must* accept Z."

"And why *must* I?"

"Because it follows *logically* from them. If A and B and C are true, Z must be true. You don't dispute *that*, I imagine?"

"If A and B and C are true, Z *must* be true," the Tortoise thoughtfully repeated. "That's *another* Hypothetical, isn't it? And, if I failed to see its truth, I might accept A and B and C, and *still* not accept Z, mightn't I?"

"You might, …So I must ask you to grant *one* more Hypothetical."

"Very good. I'm quite willing to grant it, as soon as you've written it down. We will call it:

(D) If A and B and C are true, Z must be true." …

"And at last we've got to the end of this ideal race-course! Now that you accept A and B and C and D, *of course* you accept Z."

"Do I?" said the Tortoise innocently. "Let's make that quite clear. I accept A and B and C and D. Suppose I *still* refused to accept Z?"

"Then Logic would take you by the throat, and *force* you to do it!" Achilles triumphantly replied. "Logic would tell you, 'You ca'n't help yourself. Now that you've accepted A and B and C and D, you *must* accept Z!' " (Carroll 1895, 279)

You see where this is going. The Tortoise will not accept the inference from A, B, C, and D to Z until the conditional statement explicitly linking them is included among the premises as statement E. This leads to an infinite regress of conditionals that the Tortoise must accept before he will accept the inference to Z, which implies that he will never actually accept Z. The end.

Carroll himself was coy as to his moral. He never explicitly states his point. The story ends with Achilles endlessly writing more conditional statements in his notebook and the Tortoise always insisting still more are needed.

Now, I have quoted Carroll at such length, both to include the key points of the dialogue and to illustrate the writing style he employed. I am guessing that by the time you reached the end of these excerpts, you had already formulated your own replies to the Tortoise. The first time I read the paper, I was shouting at my computer by the time I was halfway through. I found it ironical that Achilles should mention grabbing a person by the throat, since that is precisely what I wanted to do to *him* for responding so ineffectively. I had *very* definite ideas about what the Tortoise could do with his regress.

The scholarly reaction to "What the Tortoise Said to Achilles" was the reverse of the reaction to "A Logical Paradox." The latter paper caused a stir in its own time but is little discussed today. The former paper, by contrast, was largely ignored in its own time but is much discussed today.

Philosopher Pascal Engel describes the modern view of Carroll's essay like this: "[I]n the case [of Carroll's regress], everyone agrees that there is a regress, and that it is harmful, although the problem is to understand what kind of harm is done here" (Engel 2016, 85). Philosopher Corine Besson has expressed similar views: "It is admittedly very difficult to interpret exactly what Carroll's regress argument says or establishes. Its influence on philosophical logic is nonetheless immense" (Besson 2018, 505). However, the journey to that immense influence was long and strained. I consider a few of the more important scholarly responses to the regress in Section 6.3 and then offer my own views in Section 6.4.

6.3 Scholarly Responses to Carroll's Regress

The first scholar to take serious notice of Carroll's essay was Bertrand Russell, in his 1903 book *The Principles of Mathematics*. Russell sought to provide a precise statement of all fundamental logical principles used by mathematicians. While discussing certain questions that arise when reasoning with conditional statements, Russell writes:

[I]t is necessary to consider a very difficult logical problem, namely, the distinction between a proposition actually asserted, and a proposition considered merely as a complex concept. ...The principles of inference which we accepted lead to the proposition that, if p and q be propositions, then p together with "p implies q" implies q. At first sight, it might be thought that this would enable us to assert q provided p is true and implies q. But the puzzle in question [Carroll's regress] shows that this is not the case, and that, until we have some new principle, we shall only be led into an endless regress of more and more complicated implications, without ever arriving at the assertion of q. (Russell 1903, 34–35)

Russell thinks the Tortoise has a point. And if *Russell* thinks the Tortoise has a point, then that reflects very well on the Tortoise.

What is the "new principle" to which Russell refers? In his view, our system of logic must include the concept of "therefore," which is different, he says, from the concept of "implies." That is, there is a distinction between saying, "p implies q" on the one hand, and "p, therefore q" on the other. When we say "p implies q," we are not asserting p and q as individual propositions. We are only asserting a logical relationship between them. In contrast, when we say "p, therefore q," we are not asserting a proposition at all. Instead, we are making an argument. The statement "p implies q" might be true and it might be false. The utterance "p, therefore q," however, can only be valid or invalid. Russell writes:

> When we say *therefore*, we state a relation which can only hold between asserted propositions, and which thus differs from implication. Whenever *therefore* occurs, the hypothesis may be dropped, and the conclusion asserted by itself. This seems to be the first step in answering Lewis Carroll's puzzle. (Russell 1903, 35)

Russell never mentions Carroll again, thereby giving no indication of what the second step might be.

On Russell's account, Achilles ought to have said, "My dear Tortoise, you have conflated two different concepts. When we say that A and B together imply Z, we are discussing a relationship among abstract propositions. These three propositions stand in a relation of implication to one another by virtue of their abstract logical form. While this is a nice observation, it has no immediate relevance to the challenge you have posed me. *That* challenge was about being forced to accept the conclusion of a certain argument. Meeting that challenge requires invoking something more than implication relations among propositions. Indeed, the regress to which you point shows precisely why we must treat these as two separate issues. Addressing your challenge requires that we lay down principles that govern the use of the term "therefore." These principles would be things we simply accept without proof, as opposed to principles that we derive from notions that are more fundamental. You would be *forced* to accept Z only if you are willing to accept these other principles."

For many years after Russell, philosophers paid very little attention to this problem. Carroll's regress did not return to any sort of prominence until four decades later, when philosopher Gilbert Ryle mentioned it in a 1945 paper.

Ryle described the plight of a student who is made aware of various logical principles, but who is unable to use them successfully when reasoning. In the terminology of his paper, he argues that the student "knows that" certain logical principles hold, but does not "know how" to use them in practice. He writes:

> And still the pupil fails to see. And so on for ever. He accepts rules in theory, but this does not *force* him to apply them in practice. He considers reasons, but he fails to reason. (This is Lewis Carroll's puzzle in "What the Tortoise said to Achilles." I have met no successful attempt to solve it.) (Ryle 1945–1946, 6)

It is interesting that Ryle proclaimed himself unacquainted with any successful solution to the regress. In a subsequent paper, published in 1950, he offered his own solution, and his reasoning is strikingly similar to what Russell had argued almost 50 years earlier. Like Russell, Ryle argues that the Tortoise has a point and that the point involves making a distinction between "implies" and "therefore." He considers the question of how the argument "*p*, therefore *q*" might depend for its validity on the truth of the proposition "if *p*, then *q*." He suggests that most people see the relationship like this:

> An argument "*p, so q*" is always invalid unless the premiss from which "*q*" is drawn incorporates not only "*p*" but also "*if p, then q.*" "*q*" follows neither from "*if p, then q*" by itself, nor from "*p*" by itself, but only from the conjunction "*p and (if p, then q).*" (Ryle 1950, 336)

This well describes the Tortoise's reasoning. Ryle finds it unacceptable:

> But this notoriously will not do. For, suppose it did. Then a critic might ask to be satisfied that "*q*" was legitimately drawn from "*p and (if p, then q)*"; and, to be satisfied, he would have to be assured that "*if (p and [if p, then q]), then q.*" So this new hypothetical would have to be incorporated as a third component of the conjunctive premiss, and so on forever—as the Tortoise proved to Achilles. (Ryle 1950, 336)

Ryle made no explicit mention of Lewis Carroll in his paper. He merely took for granted that anyone reading his work would be familiar with the Tortoise and Achilles. (Indeed, he regards their dialogue as "notorious.")

He now summarizes his conclusion:

> The principle of an inference cannot be one of its premisses or part of its premiss. Conclusions are drawn from premisses in accordance with principles, not from premisses that embody those principles. The rules of evidence do not have to be testified to by the witness. ... [I]n saying '*q*, because *p*,'

we are not just asserting but *using* what is expressed by 'if *p* then *q*.' (Ryle 1950, 336–337)

This quickly became the canonical solution to Carroll's regress. In light of what the Tortoise said to Achilles, we must acknowledge that logical principles are separate from the arguments they underwrite. As we are steepling our fingers and deducing things in the quiet of our offices, the principles of correct inference are minding their own business somewhere else. Good fences make good neighbors.

Around this time, the correctness of Ryle's thesis, and the importance of Carroll's regress in illustrating it, came to pervade other areas of inquiry. An example is an influential book on the philosophy of science published by Stephen Toulmin in 1953. Toulmin saw an analogy between the role of natural laws in scientific reasoning and the role of inference principles in logic:

> Lewis Carroll showed in his paper, *What the Tortoise Said to Achilles*, what impossible conclusions one is led into if one treats the Principle of the Syllogism as a super-major premise, instead of as an inference-license...The conclusions about the world which scientists derive from laws of nature are not deduced from these laws, but rather drawn in accordance with them or inferred as applications of them, as our examples have shown. (Toulmin 1953, 102)

It must be said, however, that Ryle was somewhat imprecise in his argumentation. It is not clear what he means by the term "principle of an inference." The utterance "Modus ponens is a valid argument form" is certainly the expression of an inference principle. It is also a proposition, however, and there is no reason at all why it cannot be a premise in a proper deductive argument. Ryle might have been on to something, but more guidance is needed regarding what it was he meant to forbid.

Such guidance was provided by J. F. Thomson, writing in 1960:

> The proposition that such-and-such an argument is valid can itself be a premise of an argument. But it cannot be a premise in the argument to which it refers. If you want to say of some argument that it is valid you must be able to say what argument it is that you want to make this claim for. The argument must be identifiable. And the identification must be such as to allow the claim that it is logically valid to be assessed. To assess that claim we need to know what the premises are and what the conclusion is. So the premises must be identifiable independently of the claim that there are enough of them. (Thomson 1960, 101)

We should also note that Thomson was far less impressed with Carroll than were Russell and Ryle:

The extreme eccentricity of the behaviour of both of the characters may well make us wonder whether Lewis Carroll knew what he was up to in writing the story. Certainly it cannot be merely taken for granted that he intended to advance some moderately clear thesis or theses about inference but chose to do so in a veiled and cryptic way. It is just as likely that the story is the expression of a perplexity by someone who was not able to make clear to himself just why he was perplexed. (Thomson 1960, 99)

Let us now consider the merits of Thomson's idea. There still seems to be a problem. It is too blunt to say, "it cannot be a premise in the argument to which it refers." Adding premises of any sort to a valid argument cannot suddenly make it invalid. When Achilles augments his original argument by adding *C* as a premise—thereby producing the argument that *A*, *B*, and *C* imply *Z*—the resulting argument is valid. No logical harm was done by adding as a premise the assertion that *A* and *B* imply *Z*. The point is that premise *C* is redundant, in the sense that the argument was valid without it.

This was pointed out by William Bartley in 1962, though he was not specifically referring to Thomson when he did so. (Note that this is the same Bartley who discovered Carroll's lost manuscript, as we discussed in Chapter 5.) Bartley was unimpressed with the arguments made by Ryle and Toulmin:

> Although Ryle's view has been widely accepted, I know of no attempt to explain what it involves and how it is to be understood, nor of any effort to provide the view with some argumentative support. Rather, Lewis Carroll's puzzling essay, 'What the Tortoise Said to Achilles,' is regularly appealed to as if it were an authoritative and lucid demonstration of the view's correctness. (Bartley 1962, 28)

Bartley is especially critical of Ryle's assertion that it is the inclusion of principles of inference as premises in an argument that leads to infinite regress. It is here that he counters with the point I just raised—that adding premises to a valid argument cannot suddenly render it invalid. (And adding premises to an invalid argument will either make it valid or just leave it invalid.) Bartley writes:

> What leads to an infinite regress is not the use of any particular statement, be it logical rule or natural law, to strengthen or to defend an argument, whether valid or invalid, but the claim that one can thereby finally establish or prove the truth of a conclusion or the validity of an argument. (Bartley 1962, 32)

Though Bartley is critical of Ryle, and by implication, of Russell, he has arrived at a similar place. Like Ryle and Russell, he makes a distinction between the logical relationships among the premises on the one hand, and using deductively valid arguments to establish the truth of propositions on the other. You cannot prove that an argument is valid, or establish that a conclusion is true, simply by adding premises to the argument.

Bartley is strongly critical of Carroll's sloppy presentation, but nonetheless suggests the following as a point that might be drawn from the story: "Given a challenge to the validity of an argument, we give no defence by simply adding in, as another premiss, a version of the logical rule in accordance with which the argument proceeds." (Bartley 1962, 30)

Interestingly, Bartley's paper was not primarily about Carroll's regress. He discussed the regress as a side issue because it was relevant to his real interest, which involved certain claims that had been made by other philosophers regarding the nature of causal explanation in the study of history. This is another example of Carroll's regress finding relevance in fields well beyond those practiced by its original audience.

Perhaps, though, there is something bothering you about this discussion. To this point, all of the writers we have discussed have taken the view that Carroll's regress is fundamentally making a point about logical validity. Thomson bluntly stated what other writers seem to have taken for granted: "The intention of the story is, plainly enough, to raise a difficulty about the idea of valid arguments, a difficulty similar, or so Carroll implies, to Zeno's difficulty about getting to the end of a race-course" (Thomson 1960, 95). However, is it really so clear that this was Carroll's intention?

When one rereads the story, it seems that Achilles and the Tortoise actually agree on the nature of validity. The Tortoise understands perfectly that the inference from A and B to Z is logically valid. What he denies is the notion that this validity forces him to accept that Z is true. In other words, the issue is about the Tortoise's epistemic obligations, as opposed to the mechanics of what follows from what. In a 1979 paper, Barry Stroud pursued this line of inquiry. He says of the moral of Carroll's regress, "It is a point about inference and belief, and not simply about consequence and truth, or the logical relations among propositions" (Stroud 1979, 184).

In Stroud's view, the regress does not manifest itself in the infinitely long lists of premises we allegedly must add to an argument before we are compelled to accept its conclusion. Instead, it appears when we ask: What does it mean to believe a proposition Q "on the basis of" some other proposition P? He writes:

> The moral is that for every proposition or set of propositions the belief or acceptance of which is involved in someone's believing one proposition on the basis of another, there must be something else, not simply a further proposition accepted, that is responsible for the one belief's being based

on the other. It is perhaps unobjectionable for certain limited purposes to represent a cognitive subject in abstract terms as a set of propositions that are said to constitute his 'belief set'. But from Lewis Carroll's story we can conclude that no such fiction could ever represent any of a person's beliefs as based on other beliefs of his. A list of everything a person believes, accepts or acknowledges must leave it indeterminate whether any of those beliefs are based on others. (Stroud 1979, 187–188)

His argument for this conclusion is, by now, a familiar one. Even if *P* logically implies *Q*, the fact that a person believes *P* is insufficient to conclude that he also believes *Q*. He might not realize that there is any connection between *P* and *Q*, for example. There must be something that links *P* and *Q* in his mind. But if that something were a proposition (say, *R*), then we are plunged into regress. For we would now need to find a more complicated proposition (say, *S*) that links *P*, *Q*, and *R*. And so on.

In a 1995 paper, published in *Mind* as part of a celebration of the centenary of Carroll's paper, Simon Blackburn likewise pursued this theme. Blackburn expressed the point of Carroll's paper thus: "The problem [Carroll] raised can succinctly be put like this: can logic make the mind move?" (Blackburn 1995, 695). He arrived at a similar conclusion to that of Stroud: "I want to ask whether the will is under the control of fact and reason combined. I shall try to show that there is always something else, something that is not under the control of fact and reason, which has to be given as a brute extra, if deliberation is ever to end by determining the will" (Blackburn 1995, 695).

We see, then, that some scholars see in Carroll's regress a lesson about the validity of arguments, while others see instead a lesson about belief formation and inference. Still others argue that the regress has lessons for other branches of inquiry, such as the philosophy of science or the philosophy of history.

Incredibly, there is more. *Much* more.

Another line of inquiry influenced by Carroll's regress involves the normativity of logical rules. What does it mean to say we *ought* to reason in accordance with logic? When we say something like, "Having accepted *A*, you must accept *B*," what do we convey by using the word "must"?

Philosopher Pascal Engel helpfully frames the issue by distinguishing three questions:

(1) Do the two premises *A* and *B* imply the conclusion *Z*?
(2) Given that one has good reasons to believe that *A* and *B* are true, is it reasonable to believe that *Z* is true?
(3) Supposing it is reasonable for me to believe *Z* on the basis of *A* and *B*, what is supposed to move our mind to believe that *Z*? (Engel 2016, 93)

Engel writes:

> (1) is not in question, unless we revise logic strongly ... Neither is (2), if we are equipped with an appropriate justification of our rules of deduction. The point of scepticism about the force of logical reasons is that neither a positive answer to (1) nor a positive answer to (2) can yield a positive answer to (3). However willing to recognize the validity of the inference, and however well equipped with a full justification of deduction in general, this kind of sceptic will not be moved. (Engel 2016, 93)

Engel's phrasing is sloppy, since question (3) is not the sort to which you can give a positive answer, but his point seems clear enough.

Philosophers have had a good time sinking their teeth into question (3). To give some idea of the points that get raised and discussed, let us consider a 2013 paper by Jan Willem Wieland, suggestively titled "What Carroll's Tortoise Actually Proves."

Wieland discusses arguments of the following schematic form (Wieland 2013, 986):

(a) S intends to φ.
(b) S believes that φ-ing requires S to ψ.
(z) S ought to intend to ψ.

For example, if I intend to publish a paper, and if I believe that publishing a paper requires me to stay home, then I ought to intend to stay home. The problem is this: If (a) and (b) are among my beliefs, what is it, exactly, that links them to (z)? People engage in this sort of reasoning all the time, and the inference from (a) and (b) to (z) would generally be considered so obvious and natural that it hardly needs defending. The fact remains, however, that this argument is not classically valid, making it reasonable to wonder what must be added to our conceptual toolkit to justify the inference.

Wieland notes that four sorts of explanations have been suggested. It might be that the link from (a) and (b) to (z) is found in something external to the person's beliefs. That "something" might be an additional proposition she should add to her beliefs, such as "If S intends to φ, and believes that φ-ing requires S to ψ, then S ought to intend to ψ." In this case, however, we might wonder what should move her to accept *this* proposition if it is not already among her beliefs.

Alternatively, recalling the distinction made earlier between premises and rules of inference, we might find the something extra in a specific rule that she should use that instructs people to infer (z) from (a) and (b), in the same way that modus ponens instructs you to infer q from p and "if p, then q." The two remaining possibilities involve looking for the extra something among certain ideas she already accepts. These would be internal solutions, as opposed to the external solutions we just considered.

That is, there might be some proposition that she already believes, or some rule that she already accepts, that justifies the inference from (a) and (b) to (z).

All these ideas have had their defenders, and there are many pros and cons to be mooted about each. What is relevant for us is that Carroll's regress has played a central role in the discourse. Wieland (2013, 989) notes three ways in which Carroll's regress has been employed in these discussions:

1. The solutions that introduce additional premises, rather than rules, are committed to regress and hence fail.
2. The internal solutions, which introduce additional attitudes, are committed to regress and hence fail.
3. The solutions that implicitly invoke additional obligations are committed to regress and hence fail.

Wieland argues that the first two hypotheses are mistaken, their philosophical defenders notwithstanding, and that only the third is correct:

> Why should I adopt certain attitudes given certain other attitudes that I have? By many eyes, Carroll's Tortoise has something important to say about this problem. I agree. Yet, her importance does not lie where commentators usually think it lies. As I argued in this paper, the Tortoise does not demonstrate that no extra premises (rather than rules) should be introduced in our reasoning, nor that whatever is to govern our attitudes (premises or rules) should remain external to our attitudes. Rather, she shows that no solution to this problem should entail [that] our obligations to adopt certain attitudes depend on additional obligations to adopt further attitudes. (Wieland 2013, 996–997)

There are not many works of nineteenth-century logic that retain any relevance for researchers today, but Carroll's whimsical little dialogue between the Tortoise and Achilles is one of them. While no one seems to agree on the proper conclusion to draw from the story, or even on the branch of philosophy to which that conclusion belongs, it seems clear that researchers in a variety of fields have been led, inevitably, to Carroll's work. That is a far more impressive legacy than most logicians can boast of.

6.4 Does the Tortoise Have a Point?

The survey in Section 6.3 is just a small taste of the various points and counterpoints raised about Carroll's paper. Space limitations have forced me to leave out not only many important papers but also many nuances introduced by the authors whose work I *have* discussed. I hope, though, that I have provided some understanding of the major points at issue.

It is a nice observation that certain superficially plausible ways of thinking about the story lead to unacceptable regresses. However, while much of the work on Carroll's puzzle is very clever and enlightening, in my view, there is something unsatisfying about the discussion as a whole. As investigations into obscure questions in the philosophy of logic, the many papers in this area make for interesting reading. As responses to the Tortoise, however, they seem to me to miss the point.

The main issue raised by the story is not about validity, belief formation, or the normativity of logic. Instead, it is that the Tortoise has an entirely mistaken view of what logic is all about. He seems to think that logic is a dictatorial authority to whom we are answerable, as opposed to a tool to be used for solving certain sorts of problems. Too many philosophers seem willing to join him in that misapprehension.

Let us return to the argument that the Tortoise presents to Achilles:

(A) Things that are equal to the same are equal to each other.
(B) The two sides of this Triangle are things that are equal to the same.
(Z) The two sides of this Triangle are equal to each other.

At this point we should say, "The inference from A and B to Z is valid, given the principles of classical logic." When the Tortoise professes to understand perfectly the logical principles underlying the inference, but nonetheless refuses to accept Z, we should say: "The Tortoise is not reasoning in accord with classical logic." And if a reason is sought for why the Tortoise ought to align his reasoning with classical logic, then we should say: "Attempts to solve geometrical problems by methods other than those sanctioned by classical logic usually lead to undesirable results."

Done. With regard to the events in Carroll's story, there is nothing interesting to say beyond this.

When the Tortoise challenges us to force him to accept Z, we should simply decline the challenge. The only sense in which logic can force anyone to do anything is the same sense in which the rules of chess force a person to move her pieces in particular ways. If you want to play chess, then you had better move your bishops along diagonals. Any Tortoise who insisted he wanted to play chess, understood the rules, but refused to move his bishops along diagonals would just be talking nonsense. You would not try to argue with such a Tortoise. You would just nod politely and back slowly away from the conversation. So it is with logic. The inference from A and B to Z is a legal move in the game of logic, regardless of what the Tortoise thinks. The Tortoise is free to say he chooses not to reason logically, in which case we should just shrug and tell him to do what he wants.

Some philosophers find it important to convict the Tortoise of thought-crime. They want to hurl a pejorative, as though he is a naughty Tortoise for not accepting the inference to (Z). They argue that no one cares if the Tortoise declines to play chess, but if he declines to align his reasoning with logic, then he is *doing something wrong* and deserves to be criticized. There are many epithets that get thrown around to make this point: The Tortoise is "irrational." He is not living up to his "epistemic obligations." He does not understand that logic has "normative force."

Until these terms are given precise definitions, it is impossible to talk sensibly about them. But once they *are* defined with precision, all mysteries about them tend to disappear. You are welcome to devise some sharp criterion for rationality, for example, and then we can have fun deciding whether the Tortoise meets that criterion. Why we would find it interesting to have such a conversation is entirely unclear, however.

Logic has normative force only in the sense that we get good results when we apply its principles to practical problems and bad results when we do not. When Aristotle's disciples grouped his six works on logic under the title *The Organon*—meaning "the tool"—they knew what they were doing. Logic is a tool for thinking clearly. It is a very powerful tool, and one that is usefully applied in many areas of life, but it is just a tool nevertheless. The Tortoise seems to have an interest in Euclidean geometry, but he wants to undertake its study with something other than classical logic. He is guilty of nothing more than using the wrong tool for the job.

It is often said that logical truths are necessary, in the sense that they must be true in any possible world. To the extent that we are talking about purely abstract principles, that may be so. However, it is easy to imagine a world in which the set of principles we group under the heading "classical logic" is not useful for modeling our daily experiences. If the world were superintended by a malicious, omnipotent deity who deliberately acted to ensure that there is no regularity in nature, so that nothing is predictable, then it is doubtful that "logician" would ever have become a profession. The abstract principles of classical logic may express necessary truths, but it is a contingent fact of this world that they so effectively model much of our daily experience.

Which brings us to the question asked in the title of this section. Does the Tortoise have a point? The answer is no. If he came to us demanding that we force him to use a hammer to pound a nail into a piece of wood, we would not suddenly be moved to hold forth on the philosophy of hammers. We would simply have replied, "Good luck pounding the nail with any other tool." His actual demand is scarcely more complex than this, and it should be met with essentially the same response.

Readers familiar with philosophy jargon will recognize my position as broadly instrumentalist. As applied to logic, instrumentalism is roughly the view that logical systems should be assessed as useful or not useful, as opposed to correct or incorrect. It will be convenient to defer further discussion of this point to Section 12.1, where it will naturally arise again.

We have now come to the end of our discussions of Carroll's papers in logic. The scholarly consensus is that Carroll's books on logic are amusing but unimportant, while his contributions to *Mind*, especially the dialogue between the Tortoise and Achilles, represent his true contributions to the field.

The reality is precisely the reverse. Carroll's books were tremendously innovative in their presentation, which should be recognized as an important accomplishment. It was his two scholarly papers, by contrast, that were amusing but not terribly important.

6.5   *Notes and Further Reading*

Both of Carroll's contributions to *Mind* involve issues related to the proper understanding of conditional statements (also called "hypotheticals"). This problem was very much on the minds of the logicians of the time. The publication of the barbershop puzzle was preceded by extensive correspondence between Carroll and other prominent logicians, in which they argued over the correct interpretation of hypotheticals. This history is recounted by Moktefi (2008, 490–493).

There were several additional replies to the barbershop puzzle beyond what I have mentioned here. The essay by Sidgwick (1894) is very lucid. Two that are especially interesting are the papers by E.E.C. Jones (1905) and the reply to her by John Cook Wilson (1905) (who, amusingly, signed his papers simply as W.) Jones argued that the Johnson/Russell solution to the puzzle was not correct, on the grounds that "if p, then q" and "if p, then not q," taken together, do not necessarily prove that p is false. Wilson was strongly critical of her argument. The details would be too tedious to be worth discussing here, but the mere existence of the argument shows how unsettled the question of conditionals was at that time.

Wilson was a frequent sparring partner for Carroll, and they disagreed on various logical issues. With this as background, it is amusing to take note of Wilson's impatient response to the barbershop puzzle:

> It is difficult to see how this argument, which is no paradox but a paralogism, should have been taken seriously, and the fallacy in it should not have had its true character made clear nor have received the simple treatment of which it is capable. It is merely a question of language, and hardly needs the technicalities of Logic. (Wilson 1905, 292)

The term "paralogism" refers to an error made in the application of logic. Wilson was basically saying that Carroll just did not know what he was talking about.

"What the Tortoise Said to Achilles" is the subject of a vast literature well beyond the few items I mentioned in my survey. The bibliography compiled by Abeles and Moktefi (2016) is the place to start if you wish to pursue this topic. A complete survey of everything that had been said about the regress would require writing a book. That task has been undertaken by philosopher Corine Besson (2020).

In Section 6.4, I remarked that a world dominated by malicious, omnipotent deities would be one in which classical logic would not be a useful model for our experiences. We get a taste of what such a world would be like when we watch a skillful magician. The whole point of a magic trick is to show you something that seems to be impossible based on your understanding of how the world works. When we see a magic act, we know that even when we are fooled, it is ultimately just a trick, and the world returns to normal once we leave the theater. Imagine, though, that the experiences of our daily lives were like one long magic act. I am suggesting that the residents of such a world would not find classical logic to be helpful for organizing their experiences.

The question of the normativity of logic is the subject of a large literature. Surprise! Harman (1984) argued that logic simply has no normative force over our principles of reasoning. The survey article by Steinberger (2016) discusses some responses to Harman's challenges and provides a helpful bibliography.

PART III

Raymond Smullyan and Mathematical Logic

CHAPTER 7

Liars and Truthtellers

Lewis Carroll empuzzled the central principles of Aristotelian logic, thereby making them accessible to an audience that would find the textbook treatments unintelligible. In so doing, he achieved something of great importance, independent of any debates regarding his contributions to scholarship.

Raymond Smullyan achieved something similar for propositional logic, the central principles of which are explained in Section 7.1. His preferred mode of empuzzlement was to imagine an island on which every inhabitant is either a knight or a knave. Knights always tell the truth, while knaves always lie. The two tribes are visually indistinguishable, meaning that the only way of determining who is who is to ferret out the logical consequences of their often cryptic remarks.

We had a taste of this kind of puzzle in Chapter 2, in Puzzles 11 and 12. You might wish to go back and review those puzzles before moving on.

The main business of the present chapter is to provide a small selection of Smullyan's puzzles for your enjoyment. Most of these puzzles will involve knights and knaves, but others will involve liars and truthtellers of other sorts.

First, however, we must discuss the basic principles of propositional logic.

7.1 Propositional Logic

Aristotle studied the logic of class inclusion. He noted that certain pairs of categorical statements force you to certain conclusions, on pain of otherwise being thought irrational. For him, the fundamental units of logical analysis were classes of objects.

However, categorical statements are not the only sort that impose logical burdens on the people who hear them. Collections of ordinary, declarative sentences can impose such burdens as well. If a truthful person tells you that they will be serving either cookies or cake for dessert, and later they tell you they have decided not to serve cake, then you are forced to conclude that they will be serving cookies. And just as Aristotle sought to formalize the logical burdens imposed by categorical statements, so, too, we might hope to formalize the logical burdens imposed by declarative sentences.

It is from such considerations that propositional logic is born. As the name suggests, the fundamental units in propositional logic are propositions. In nearly all situations, you will not go wrong by equating propositions with declarative sentences, and we could as easily speak of "sentential logic" as "propositional logic." However, in Section 1.1, we noted the reasons for distinguishing between sentences and propositions, and we shall use the latter term for this discussion. (The distinction will again become relevant in Section 16.4.)

To begin, notice that if P is any proposition, there seems to be another proposition whose meaning is "P is false." This other proposition is referred to as the *negation* of P and is denoted by ¬P. For example, if P is "Grass is green," then ¬P is "Grass is not green." If P is "The Moon is made of green cheese," then ¬P is "The Moon is not made of green cheese." For simple propositions, the negation is typically formed by inserting "not" into the sentence at some appropriate place. For more complex propositions, it might be more difficult to work out precisely what the negation is.

It would seem there are a few basic principles we can lay down even at this early stage. For instance, between any proposition and its negation, exactly one of them is true. Either the Moon is made of green cheese or it is not, and it cannot simultaneously be made of green cheese and not be made of green cheese. I do not mean to say that I can prove this assertion—that exactly one of P and ¬P is true—in terms of something simpler. Instead, I am suggesting that this principle seems to capture everyone's intuitive understanding of how propositions relate to their negations. The notion that at least one of a proposition and its negation is true is known as the *law of the excluded middle*. The notion that a statement and its negation cannot both be true simultaneously is known as the *law of noncontradiction*.

There are already problems, however. In everyday life, some sentences are inherently vague, in that any attempt at precisification would just be arbitrary. You might not want to assign a truth value either to the sentence or its negation. In other cases, a sentence is true in one sense and not in another. Is it not reasonable to say, for example, that a teenager is in some ways like a child and in some ways not like a child? However, it *does* seem to be the case that many propositions are precise, and our basic principles apply very well to them.

Let us now consider some ways in which simple propositions can be joined to form more complex propositions. One way this can be done is to place the word "and" between them. In routine discourse, this is generally understood to mean that both propositions must be true individually.

When I was in high school, one of my history teachers liked taking advantage of this fact to produce devious true/false questions for his exams. He would make a statement like, "Thomas Jefferson was the third President of the United States, and he was once the Secretary of State, and he served as the Governor of Virginia, and he was born in 1743," and you would have to know that all four parts are true (they are) in order to determine that the whole statement was true. If you were uncertain, the best course was to guess false, since only one of the component propositions needed to be false to make the whole thing false.

Let us use the symbol $\wedge$ to denote "and." Then we can express our intuition formally by saying that if P and Q are two propositions, then the proposition $P \wedge Q$ is true only when P and Q are true individually, and it is false otherwise. A statement that employs the connective $\wedge$ is commonly referred to as a *conjunction*.

Table 7.1 summarizes our conventions regarding negations and conjunctions.

Two other connectives are commonly used. Sometimes we join two propositions with "or," which is generally taken to mean that at least one of the individual propositions is true. Here again, however, care is needed. In everyday discourse, the connective "or" has two different meanings. Sometimes we mean simply that at least one of the individual propositions is true. If I say, "At the party, I will eat cookies, or I will eat cake," I am certainly not ruling out the possibility of eating both. If I am watching a sporting event, I might say, "Either team A will win, or team B will win," but I certainly am not leaving open the possibility that both will win.

TABLE 7.1.
The standard truth tables for negation and conjunction.

P	¬P		P	Q	P∧Q
T	F		T	T	T
F	T		T	F	F
			F	T	F
			F	F	F

Table 7.2.
The standard truth table
for disjunctions.

P	Q	P ∨ Q
T	T	T
T	F	T
F	T	T
F	F	F

Let us use the symbol ∨ to represent the logician's version of "or." A proposition employing this connective is called a *disjunction*. The usual convention is to interpret ∨ to be the inclusive or—it indicates that at least one of the individual propositions is true. It can only be false if both individual propositions are false. This is shown in Table 7.2.

Finally, we must consider conditional statements, which employ the connective "if–then." For example, in mathematics, I might say something like, "If a, b, and c represent the lengths of the sides of a right triangle, with c denoting the length of the hypotenuse, then $a^2 + b^2 = c^2$." In everyday life, I am more likely to say something like, "If it rains tomorrow, then I will go to the movies." Plainly, sentences employing this construction impose logical demands on those who hear them. But it is not so simple to determine what those demands are.

Let us use the symbol → to represent conditional statements. Thus, you should interpret "P → Q" to mean, "If P, then Q." In such a statement, the proposition P is said to be the *antecedent* of the conditional, and the proposition Q is said to be the *consequent*.

It seems clear that if the antecedent of a conditional is true and the consequent is false, then the entire conditional statement should be evaluated as false. For example, if you say, "If it rains, then we will go to the movies," and it subsequently rains and you do not go to the movies, then everyone would agree that you said something false.

The problem is how to assess the other possibilities. Toward that end, it will be helpful to distinguish between conditional statements in general and the subset of conditional statements that arise as mathematical theorems. The distinction is that mathematical theorems describe logical entailments, whereas conditionals in general do not. That is, a mathematical theorem asserts that if the antecedent is true, then it is logically necessary that the consequent be true as well, but this is not the case in general.

Now, what should we think of a conditional statement in which the antecedent and consequent are both true? Should we conclude that the entire conditional statement is true? If we are thinking of conditionals in general, then this seems premature, since we generally expect that in a true conditional, the antecedent is relevant in some way to the consequent. A statement such as "If snow is white, then grass is green" does not appear to be true, even though both parts are true. This issue does not arise for mathematical theorems, however. If the antecedent logically implies the consequent, then the relevance of the two parts is built into the proposition, so to speak. Thus, for assertions of logical entailment, it does seem reasonable to judge the whole conditional to be true when there is no scenario in which the antecedent is true and the consequent is false.

What about cases when the antecedent is false? For conditionals in general, it seems strange to assign any truth values at all to these cases. If I say, "If it rains tomorrow, then we will go to the movies," and then it does not rain, we would probably just shrug and move on. We would say the conditions for assessing the truth of the statement never arose. In mathematics, however, the situation is different. In that context, it is often very convenient to speak of a conditional being "vacuously true," meaning that it is true simply because there is no instance where the antecedent is true. After all, if I assert "If P, then Q," but P is never true, then I have not actually lied to you.

The convention in propositional logic is to define the connective $\to$ in a way that makes perfect sense for mathematical theorems, even though this approach does not faithfully represent every aspect of conditional statements that arise in normal discourse. Thus, we say that a conditional statement is false when the antecedent is true and the consequent is false, but true in every other case. This is represented in Table 7.3. We can summarize Table 7.3 by saying that

$$P \to Q \equiv \neg P \vee Q,$$

where "$\equiv$" means "is equivalent to." When you assert a conditional statement, at a minimum you are saying that it is impossible for the antecedent

TABLE 7.3.
The standard truth table
for implications.

P	Q	P → Q
T	T	T
T	F	F
F	T	T
F	F	T

TABLE 7.4.
The standard truth table
for biconditionals.

P	Q	P ↔ Q
T	T	T
T	F	F
F	T	F
F	F	T

to be true while the consequent is false. In mathematical discourse, we take the view that nothing more than this is being asserted.

It is customary to refer to this account of conditional statements as the *material conditional*, or as *material implication*. The idea is that the truth of the conditional depends only on the subject matter of the individual propositions and not on any relation of causality or relevance between the two parts. It is also common to say that this account is *truth-functional*, in the sense that the truth of the conditional statement is purely a function of the truth of the individual parts.

There is one further connective that is sometimes useful, even though, strictly speaking, it is redundant, given the connectives we have already defined. I am referring to "if and only if." Propositions employing this connective are typically referred to as *biconditionals*, because they are really just a shorthand notation for two separate conditional statements. The symbol for this connective is the two-headed arrow ↔, and we use the following equivalence to define it:

$$P \leftrightarrow Q \equiv (P \rightarrow Q) \wedge (Q \rightarrow P).$$

In other words, when we assert, "P ↔ Q," we are saying, "P implies Q, and Q implies P." Using this equivalence, Table 7.4 defines the truth table for biconditionals. Thus, a biconditional statement is true precisely when the two parts individually have the same truth values.

You can think of the foregoing definitions and truth tables as the rules of the game for the puzzles to come.

7.2 A Knight / Knave Primer

Let us work through a few representative examples of knight/knave puzzles, to establish some basic principles and solution methods. In this section, solutions will be presented right after the puzzles themselves.

As a warm-up, we start with a variation on Puzzle 12.

Puzzle 23. *You meet three people, A, B, and C. You ask A: "How many knights are among you?" A replies with an unintelligible mumble. So you ask B what A said. B replies: "A said there is exactly one knight among us." At this point C interrupts and says, "Don't believe B! He's lying!" What can you determine about the three people? (Smullyan 1978, 20–21)*

Solution: The key observation is that since B and C contradict each other, one must be a knight and the other must be a knave. We now ask whether A could really have said what B says he said. If A is a knight, then that would make two knights among them, implying that A had lied. This is impossible. If A is a knave, then that would imply there really is exactly one knight among them, implying that A had told the truth. This is again impossible. This shows that A could not possibly have said there is exactly one knight among the three, implying that B was lying. So B is a knave, which implies that C is a knight. However, we cannot say anything about A. □

Here are two general principles that are often helpful for solving knight/knave puzzles:

1. No islander can claim to be a knave.
2. If two islanders contradict each other, then they must be of different types.

Keep these principles in mind while solving the next puzzle.

Puzzle 24. *You meet three natives, A, B, and C, who make the following statements:*

> A : *Exactly one of us is a knave.*
>
> B : *Exactly two of us are knaves.*
>
> C : *All of us are knaves.*

What type is each? (Smullyan 2009, 5)

Solution: A knight could not possibly have made the statement attributed to C, since that would be tantamount to a knight claiming to be a knave. So, C must be a knave. Moreover, A and B contradict each other, implying that one is a knight and the other is a knave. It follows that there must be two knaves and one knight among the three. That implies that B is the knight and A and C are knaves. □

Let us try one more before really getting down to business. For this next puzzle, we will say that two islanders are the same type if they are both knights or both knaves.

Puzzle 25. *Again there are three natives, referred to as A, B, and C. This time, only A and B speak. They make the following statements:*

> *A : B is a knave.*
>
> *B : A and C are of the same type.*

What can we conclude about C? (Smullyan 1978, 22)

Solution: To begin, let us ask what would follow if we assumed that A is a knight. In that case, it would be true that B is a knave. That would imply B's statement is false, which would immediately imply that A and C are not of the same type. It would follow that C is a knave. This scenario is possible, since it does not lead us to a contradiction. We have shown that if A is a knight, then C is a knave.

What if we assumed instead that A is a knave? Then his statement is false, implying that B is actually a knight. That would force B's statement to be true. So, A and C really are of the same type. Since A is a knave in this scenario, we would again conclude that C is a knave.

Thus we have no way of knowing the types of A and B. But we *can* be certain that C is a knave. □

Now that you have seen some of the thought processes underlying these puzzles, have a go at the sampling presented in the succeeding sections. The solutions to all puzzles from this point on will appear at the end of the chapter. Good luck!

7.3 A Selection of Knight / Knave Puzzles

Let us begin with a few warm-up puzzles.

Puzzle 26. *You meet two people, A and B. Suppose A said, "Either I am a knave or B is a knight." What would you conclude about A and B? Suppose instead A said, "I am a knave, but B isn't." What are A and B this time? Finally, suppose I told you that A said, "Either I am a knave or 2 + 2 = 5." What would you conclude? (Smullyan 1978, 21–22)*

Puzzle 27. *This time you are attending a trial on the island. Two witnesses, A and B, make the following statements about the person on trial:*

> *A : If B is a knight, then the defendant is innocent.*
>
> *B : If A is a knave, then the defendant is innocent.*

Is the defendant guilty? (Smullyan 2009, 58)

In addition to knights and knaves, Smullyan populated his islands with people who sometimes lied and sometimes told the truth. Since this is pretty much the way normal people behave, Smullyan referred to such people as normals.

So, in the puzzles that follow, we assume the islands to be populated with knights who always tell the truth, knaves who always lie, and normals who sometimes lie and sometimes tell the truth.

Puzzle 28. *We are given three people, again referred to as A, B, and C. One of them is a knight, one is a knave, and one is a normal. They make the following statements:*

> A : *I am normal.*
>
> B : *That is true.*
>
> C : *I am not normal.*

Determine the types of all three people. (Smullyan 1978, 23–24)

It was commonplace on the island to say that knights were of the highest rank, knaves were of the lowest rank, and normals occupied the middle. That will be useful in the next two puzzles.

Puzzle 29. *You meet A and B while visiting the island. They say:*

> A : *I am of lower rank than B.*
>
> B : *That's not true!*

What can you conclude from these statements? (Smullyan 1978, 24–25)

Puzzle 30. *This time you meet A, B, and C, and you know that one is a knight, one is a knave, and one is a normal. The first two natives make the following statements:*

> A : *B is of higher rank than A.*
>
> B : *C is of higher rank than B.*

At this point, you ask C, "Who has higher rank, A or B?" What answer does C give? (Smullyan 1978, 25)

7.4 Sane or Mad?

Lying and truth telling are intimately related to belief. When you set out to tell the truth, you can only say what you believe to be the truth, and when you lie, you can only say what you believe to be false. We shall say that a person is sane if all of her beliefs are accurate, and mad or insane if all of her beliefs are inaccurate. These definitions lead to intriguing possibilities for puzzles.

For example, imagine that the residents of a particular asylum consist entirely of patients and doctors. Moreover, everyone in the asylum is either sane or mad. (For this puzzle, there are no knights or knaves.) We assume that everyone strives to tell you the truth, in the sense that they accurately report what they believe. Suppose a resident in such an asylum said to you, "I am not a sane doctor." In response, we could reason like this: If the person is actually insane, then he could not accurately believe he is not a sane doctor. Moreover, if he is sane and a doctor, then he would not falsely believe that he is not a sane doctor. That would mean the resident must be a sane patient.

Imagine that another resident said to you, "I am an insane patient." This time you would reason that if the resident were sane, he would not falsely believe himself to be insane, and if he were an insane patient, he would not accurately believe that he is. Therefore, he must be an insane doctor.

The next two puzzles are based on this setup.

Puzzle 31. *One day, Inspector Craig of Scotland Yard was sent to the asylum to investigate. He separately interviewed two residents, whose last names were Jones and Smith.*

"Tell me," Craig asked Jones, "what do you know about Mr. Smith?"

"You should call him Doctor Smith," replied Jones. "He is a doctor on our staff."

Later, Craig met Smith and asked, "What do you know about Jones? Is he a patient or a doctor?"

"He is a patient," replied Smith.

The inspector mulled over the situation for a while and then realized that there was indeed something wrong with this asylum: Either one of the doctors was insane, hence shouldn't be working there, or, worse still, one of the patients was sane and shouldn't be there at all.

How did Craig know this? (Smullyan 1982b, 29–30).

Puzzle 32. *In a different asylum, Inspector Craig again met two inhabitants, A and B. A believed that B was mad, and B believed that A was a doctor. Craig then took measures to have one of the two removed. Which one, and why? (Smullyan 1982b, 31)*

The difficulties multiply when we combine the notions of sane and insane people with the notions of knights and knaves. Consider the problem of distinguishing a sane knight from a mad knave. A sane knight will want to tell the truth, and since all of his beliefs are correct, he will be successful in that goal. A mad knave wants to lie, but since all of his beliefs are false, he will inadvertently tell the truth after all. Of course, a similar problem ensues when trying to distinguish a mad knight from a sane knave.

Let us consider two puzzles about knight/knave islands on which every resident is either sane or insane (but not both!).

Puzzle 33. *You meet two natives, named Ceg and Fek. It was known that one was a knight and one was a knave. It was also known that one was sane and the other was mad. Ceg said that Fek is mad, and Fek said that Ceg was a knave. What type was each? (Smullyan 2009, 32–33)*

Puzzle 34. *One day you meet two natives, named Bek and Drog. They make the following statements:*

> *Bek : Drog is mad.*
>
> *Drog : Bek is sane.*
>
> *Bek : Drog is a knight.*
>
> *Drog : Bek is a knave.*

What can you conclude about Bek and Drog? (Smullyan 2009, 33)

7.5 The Lady or the Tiger?

In the following four puzzles, imagine that a king has decided to test his prisoners for their ability to reason. Each prisoner is confronted with a collection of doors. Each door conceals either a lady or a tiger. The prisoner is provided with certain information regarding the truth of the signs and the distribution of ladies and tigers. He then has to select a door that conceals a lady.

Puzzle 35. *The king showed the prisoner the signs displayed in Figure 7.1. The prisoner asked, "Are the signs true?" The king replied, "They are*

At least one of these rooms contains a lady.	A tiger is in the other room.

Figure 7.1. The signs for Puzzle 35.

either both true or both false." Which room should the prisoner pick? (Smullyan 1982b, 16)

Puzzle 36. *"There are no signs above the doors!" exclaimed the prisoner.*
"Quite true," said the king. "The signs were just made, and I haven't had time to put them up yet."
"Then how do you expect me to choose?" demanded the prisoner.
"Well, here are the signs," replied the king (see Figure 7.2).
"That's all well and good," said the prisoner anxiously, "but which sign goes on which door?"
The king thought awhile. "I needn't tell you," he said. "You can solve this problem without that information. Only remember, of course," he added, "that a lady in the left hand room means the sign which should be on that door is true, and a tiger in it means the sign should be false, and that the reverse is true for the right hand room."
What is the solution? (Smullyan 1982b, 19)

The next day, seeking to make things more difficult for the prisoners, the king added a third door to the mix.

Puzzle 37. *The first prisoner to confront the three doors was shown the signs in Figure 7.3. The king told the prisoner that at most one of the three signs is true. Moreover, one of the rooms contained a lady, but the other two contained tigers. Which room contained the lady?*

Puzzle 38. *The second prisoner to confront the three doors was shown the signs in Figure 7.4. Again there was one lady and two tigers. The sign on the room containing the lady was true. At least one of the other two signs was false. Again, where is the lady? (Ibid, 20–21)*

> This room
> contains
> a tiger.

> Both rooms
> contain tigers.

Figure 7.2. The signs for Puzzle 36.

Figure 7.3. The signs for Puzzle 37.

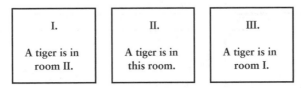

Figure 7.4. The signs for Puzzle 38.

7.6 Some Unusual Knights and Knaves

As we have seen, the typical knight/knave puzzle involves groups of people of unknown types engaging in a dialogue, at the end of which you must determine who is who. Over the years, Smullyan devised puzzles that depart in some way from this general formula. The puzzles in this section are representative of some of his formula breakers.

Puzzle 39. *One day I was visiting the island and met one of the natives. "Are you a knight or a knave?" I asked him. He angrily replied, "I refuse to tell you!" Then he stormed off. That is the last time I ever saw him or heard of him. Was he a knight or a knave? (Smullyan 1985, 43)*

Puzzle 40. *On my final visit to the island, a native said to me, "This is not the first time I have said what I am now saying." What would you conclude? (Smullyan 1985, 44)*

Puzzle 41. *A visitor to the island encounters a native. The visitor knows for a certainty that the native's name is either Paul or Saul, but he does not know which. When he asks the native for his name, the native replies, "My name is Saul." It is not possible from this for the visitor to determine either the native's name or whether he is a knight or knave. But he can answer these questions with a high probability of being correct. How is this possible? (Smullyan 2009, 5)*

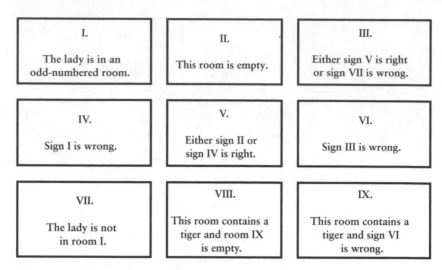

Figure 7.5. The signs for Puzzle 42.

7.7 Two Elaborate Puzzles

I close this chapter with two puzzles that are more complex than what we have seen. The first is set in the same world as the puzzles in Section 7.5.

Puzzle 42. *The next prisoner was confronted with nine doors. One of the nine contained a lady, while each of the other eight either contained a tiger or were empty. The sign on the room containing the lady is true, the signs on all of the rooms containing tigers are false, and the signs on the empty rooms could be either true or false. The signs are shown in Figure 7.5. The prisoner thought for a while. "The problem is unsolvable!" he exclaimed angrily. "That's not fair!"*

"I know," laughed the king.

"Very funny!" replied the prisoner. "Come on, now, at least give me a decent clue: is room eight empty or not?"

The king was decent enough to tell him whether room VIII was empty or not, and the prisoner was then able to deduce where the lady was.

So, which room contained the lady? (Smullyan 1982b, 22–23)

This final puzzle appeared in Smullyan's book *Alice in Puzzle-Land*, in which all puzzles are based on characters from Lewis Carroll's *Alice* books. The setup for this puzzle is that a plate of tarts has been stolen, and it is known for a fact that the thief was either the Gryphon or the Mock Turtle, but not both. The puzzle is based on what happened at the trial.

Puzzle 43. *It seemed that no further progress could be made, but quite suddenly a whole medley of witnesses came up, making various statements.*

"The Gryphon never stole the tarts," said the Duchess.

"But he has stolen other things in the past," said the Cook.

"The Mock Turtle has stolen things in the past," said the Caterpillar.

"The Cook and the Cheshire Cat are both right," said the March Hare.

"The Cook and the Caterpillar are both right," said the Dormouse.

"Either the Cheshire Cat or the Caterpillar is right—and maybe both," said the Hatter.

"Either the March Hare or the Dormouse is right—and maybe both," said Bill the Lizard.

"The Cook and the Hatter are both right," said the Knave of Hearts.

"Bill the Lizard is right and the Knave of Hearts is wrong," said the White Rabbit.

There was dead silence.

"All this proves nothing!" roared the King. "Just words, words, words—all useless words!"

"Not so useless, your Majesty," said Alice, rising from the jury. "It so happens that the White Rabbit and the Duchess made statements which are both true or both false."

All eyes turned eagerly to Alice. Now, everyone knew that Alice makes only true statements, and subsequent investigation showed that this statement was no exception. Moreover, this statement solved the entire mystery.

Who stole the tarts? (Smullyan 1982a, 16–19)

7.8 Notes and Further Reading

The definitions for the logical connectives ¬, ∧, ∨, and → certainly capture important aspects of how their English counterparts—not, and, or, and if–then, respectively—are used. It can also be interesting to ask, however, what aspects of natural language have been lost in our formalization. For example, in formal logic, the propositions P ∧ Q and Q ∧ P are treated as equivalent. In normal conversation, this assumption is not always warranted, as sometimes the order of the conjuncts affects our understanding of the sentence. For example, "I ate chicken soup and felt sick," does not give the same impression as "I felt sick and ate chicken soup." The paper by Bach (2002), from which this example is drawn, is a good reference for this topic.

The truth-functional account of conditional (if–then) statements, which I have previously labeled "material implication," seems to work well for mathematical theorems, as I have described. I have also noted that it does not work as a general account of conditional statements, since in normal conversation, we typically assume the antecedent and consequent are relevant to each other. Other problems arise from the so-called "paradoxes of material implication." For example, given our definitions, we have that the proposition

$$(P \wedge Q) \rightarrow R$$

implies that

$$(P \rightarrow R) \vee (Q \rightarrow R).$$

This can be checked readily, if somewhat tediously, by constructing a truth table. Expressed more plainly, the propositions say that if two things together imply a third, then one of the two things by itself implies the third. It is not hard to devise counterexamples: "If my bicycle has a front wheel and a rear wheel, then I can ride it. This implies either that if I have a front wheel, then I can ride it, or if I have a rear wheel, then I can ride it."

There are other instances like this, by which I mean inferences that are formally valid but can lead to absurd conclusions in normal conversation. Priest (2008b, 1–19) and Read (1995, 64–95) provide readable discussions of some of the issues surrounding material implication. For a very deep discussion of this topic, try the book by Bennett (2003).

I have used the symbol $\rightarrow$ for material implication. This is standard in mathematics textbooks, but logicians generally prefer the symbol $\supset$. They would write $P \supset Q$ to mean precisely what I mean when I write $P \rightarrow Q$. They reserve $\rightarrow$ for the stronger notion of logical implication. For our purposes, however, this distinction is not important.

In my definition of the logical connective $\leftrightarrow$, I mentioned that it was redundant, since it is defined entirely in terms of the previously defined connectives $\rightarrow$ and $\wedge$. That is, a biconditional statement is just a conjunction of two conditional statements. I also mentioned that $\rightarrow$ is defined entirely in terms of $\neg$ and $\vee$, which makes $\rightarrow$ redundant. (Recall that $P \rightarrow Q$ is equivalent to $\neg P \vee Q$.)

It is interesting to note that $\vee$ is similarly redundant, in the sense that we have the following equivalence:

$$P \vee Q \equiv \neg(\neg P \wedge \neg Q).$$

This shows that we really only need the connectives $\neg$ and $\wedge$, since the other connectives we care about can be defined in terms of just these two. It would not be very convenient to do logic this way, however.

With two propositions, there are 16 ways to fill in the last column of the truth table, if we are just assigning Ts and Fs randomly. It can be shown, though it is tedious to do so, that all 16 can be expressed only with combinations of ¬ and ∧. Even more interesting is that it is possible, in principle, to make do with only a single connective, denoted by ↑. This connective is known as the Sheffer stroke, since it was introduced by mathematician Henry Sheffer (1913). We define P ↑ Q to be true precisely when at least one of P and Q is false. We now have the equivalences

$$\neg P \equiv P \uparrow P,$$

$$P \wedge Q \equiv (P \uparrow Q) \uparrow (P \uparrow Q).$$

Since everything else is definable in terms of ¬ and ∧, we have shown that, strictly speaking, only one connective is necessary.

Finally, fans of Carroll's *Alice in Wonderland* will recognize Puzzle 43 as having been inspired by Chapter 11 of that book, which also centers on a trial about some stolen tarts.

7.9 Solutions

Puzzles 23–25 are solved in Section 7.2.

Puzzle 26.

- Suppose A says, "Either I am a knave or B is a knight." If A really were a knave, then the first part of the disjunction is true, which would make the whole statement true. This is a contradiction. So A is a knight. Since the whole statement must be true, but the first part is false, the second part must be true. So B is also a knight.
- This time A says, "I am a knave, but B isn't." This is equivalent to the statement, "I am a knave, and B is a knight." If A is a knight, then the first part of his conjunction is false, which is already a contradiction. So A is a knave, and the conjunction must be false. Since the first part of the conjunction is true, the second part must be false. So in this case, A and B are both knaves.
- Neither a knight nor a knave could make the statement, "Either I am a knave, or $2 + 2 = 5$." If A is a knight, then the conjunction is false (since both parts are false), while if A is a knave the conjunction is true (since the first part is true). Either way, we have reached a contradiction. So, you would conclude that I was pulling your leg.

Puzzle 27. Recall that an if-then statement can only be false when the antecedent is true and the consequent is false. Suppose that A is a knave. Then it must be true that B is a knight and also that the defendant is guilty. But in this scenario, the antecedent of B's statement is true. Since B is a knight, this implies the consequent must be true as well. But this would imply that the defendant is innocent after all, which is a contradiction. Now assume that B is a knave. In this case, it must be true that A is a knave and that the defendant is guilty. But the antecedent of A's statement is true in this scenario, which implies the whole if-then statement is true as well. But this is impossible when A is a knave.

Therefore, the only consistent scenario is that both witnesses are knights, and that the defendant is innocent.

Puzzle 28. First note that *A* is not a knight, since in that case, his statement would be false. If *A* is the normal, then his statement is seen to be true. In this case, *B*'s statement is also true. Since *A* is the normal in this scenario, that would make *B* the knight. Then *C* would have to be the knave, which is impossible, since his assertion that he is not normal is now seen to be true. It follows that *A* is the knave. That means *B* is a normal who is speaking falsely, and *C* must be the knight.

Puzzle 29. Neither a knight nor a knave could claim to be of lower rank than anyone else, since a knight who said that would be lying, while a knave who said that would be telling the truth. It follows that A is a normal. Now, if B is a knight, then A really is of lower rank than B, which makes B's statement false, which is impossible. Likewise, if B is a knave, then A's statement is false. But this would imply that B's statement was true, which is again a contradiction. We conclude that A and B are both normals, and that both are lying.

Puzzle 30. Suppose first that A is the knight. Then his statement is true, which implies that B is the normal and C is the knave. In this scenario, B's statement is false, which is consistent with his being a normal. So, the truth is that A is of higher rank than B. Asked about this, C, a knave, will lie and claim that B is of higher rank.

Now suppose that A is the knave. Since his statement must now be false, we would have that B is the normal and C is the knight. This scenario is also consistent. So this time, the truth is that B is of higher rank than A. Asked about this, C, a knight, will tell the truth and say that C is of higher rank.

Finally, what if A is the normal? That would imply that if B is the knight, then C is the knave. But this would make B's statement false, which is a contradiction. Alternatively, if B is the knave, then C is the knight. But in this case, B's statement is true, which is another contradiction. Therefore, it is not possible for A to be the normal.

We conclude that C will definitely claim that B is of higher rank than A. We can also conclude that B is the normal, even though we were not asked about this.

Puzzle 31. Suppose Jones is a patient. Then Smith's assertion that Jones is a patient is correct, implying that Smith is sane. If Smith is a patient, then we have found a sane patient. So, assume instead that he is a doctor. In this case, Jones's assertion that Smith is a doctor is correct, implying that he is sane. So, again, we would have a sane patient.

Now suppose Jones is a doctor. In this case, Smith's assertion is false, implying that he is mad. If he is a doctor, then we have found a mad doctor. So, assume that he is a patient. But in this case, Jones's assertion is false, implying that he is mad. So, again, we have found a mad doctor.

Puzzle 32. Resident A should be removed on the grounds that he is either a sane patient or a mad doctor (though we cannot know which). Suppose A is sane. Then his belief that B is mad is correct. That means B's belief that A is a doctor is incorrect, implying that A is a patient. In this case, A is a sane patient.

Now suppose A is mad. In this case, his belief that B is mad is wrong, implying that B is sane. But then B's belief that A is a doctor is correct, implying that A is a mad doctor.

Nothing can be concluded about B, however.

Puzzle 33. Suppose that Ceg is the knave. Then Fek is the knight. Moreover, Fek's statement is true in this scenario, implying that he is a sane knight. That implies that Ceg is mad. But a mad knave makes true statements, while Ceg's statement is false in this scenario. This is a contradiction. We have shown that Ceg is not the knave.

Therefore, Ceg is the knight, and Fek is the knave. Fek's statement is false in this scenario, implying that Fek is a sane knave. That makes Ceg a mad knight.

Puzzle 34. First note that Bek's statements are either both true or both false. If both are true, then Drog is a mad knight. If both are false, then Drog is a sane knave. In either case, both of Drog's statements are false. That means that Bek must actually be a mad knight. Since this implies that both of Bek's statements are false, we conclude that Drog is a sane knave.

Puzzle 35. It cannot be that both signs are false. To see this, notice that if the first sign is false, then neither room contains a lady. This implies that both rooms contain tigers. But this immediately makes the second sign true. It follows that both signs must be true. Since the first sign is true, there must be a lady in one of the rooms, and since the second sign is true, there must be a tiger in the first room. So, the prisoner should open the second door.

Puzzle 36. Suppose the sign that reads, "This room contains a tiger" were placed on the left hand door. Then, if this room actually contained a lady, the sign would be false, and if this room actually contained a tiger, then it would be true. Either assumption contradicts the king's rules.

It follows that "This room contains a tiger" should be on the right-hand door, and "Both rooms contain tigers" should be on the left-hand door. Now, if there is actually a lady in the left-hand room, then its sign would be false, which is a contradiction. Therefore, there is a tiger in the left-hand room. But since the sign's statement must be false, there cannot also be a tiger in the other room. It follows that there is a lady in the right-hand room. (And notice that in this case, the sign on the right-hand door is false, which is consistent with the king's rules.)

Puzzle 37. Notice that the signs on rooms II and III contradict each other. Consequently, one of the signs is true and the other is false. Since at most one the three signs is true, we conclude that the sign on room I is false. This implies that the lady is in room I.

Puzzle 38. The lady cannot be in room II, since that would make the sign on the door false. If the lady is in room III, then there are tigers in rooms I and II. But this would imply that all three signs were true, contrary to the king's instructions. We conclude that the lady must be in room I.

Puzzle 39. The native said he refused to tell me whether he was a knight or a knave. And, in fact, he did refuse to tell me! So he must have been a knight after all. Note that there is no paradox implied by this conclusion. The native never said he would take no action from which it would be possible for me to deduce whether he was a knight or a knave. He merely said he would not tell me which he was.

Puzzle 40. The native must be a knave. One way to see this is to consider that there is some minimum amount of time required for the native to utter that sentence. Let us say a person would require at least one second to utter the sentence in a manner that is comprehensible to the people he is speaking to. Well, since any person has only been alive for a finite number of seconds, he could only have uttered this sentence a finite number of times. In that case, there must have been a first time that he uttered the sentence, and when he said it *that* time he was lying. So, he is a knave.

Puzzle 41. Suppose the native is a knave. Then his name is actually Paul and not Saul. Now, when a knave named Paul is asked for his name, he will reply with any name other than Paul. For that scenario to be correct in this puzzle, it would have to be that of all the phony names Paul could possibly have given, he happened to choose the only one you knew to be an actual possibility. You cannot completely rule this out, of course, but it seems far more likely that the native is a knight whose name is Paul.

Puzzle 42. The first observation is that the king must have told the prisoner that room VIII is not empty. Otherwise, since the signs on empty rooms could be either true or false, the prisoner could not have received any useful information from what the king told him.

The prisoner now reasons as follows: "It is impossible for the lady to be in room VIII, for if she were, sign VIII would be true, but the sign says a tiger is in the room, which would be a contradiction. Therefore, room VIII does not contain the lady. Also, room VIII is not empty; therefore, it must contain a tiger. Since it contains a tiger, the sign is false. Now, if room IX is empty, then sign VIII would be true; therefore, room IX cannot be empty.

"So, room IX is also not empty. It cannot contain the lady, or the sign would be true, which would mean that the room contains a tiger; this means sign IX is false. If sign VI is really wrong, then sign IX would be true, which is not possible. Therefore, sign VI is right."

"Since sign VI is right, it follows that sign III is wrong. The only way sign III can be wrong is that sign V is wrong and sign VII is right. Since sign V is wrong, it follows that sign II and sign IV are both wrong. Since sign IV is wrong, it follows that sign I is must be right."

Now we know which signs are right and which signs are wrong:

I: Right	II: Wrong	III: Wrong
IV: Wrong	V: Wrong	VI: Right
VII: Right	VIII: Wrong	IX: Wrong

We know that the lady is in either room I, room VI, or room VII, since the others all have false signs. Since sign I is right, the lady cannot be in room VI. Since sign VII is right, the lady can't be in room I. Therefore, the lady is in room VII.

Puzzle 43. The White Rabbit claimed that Bill the Lizard was right and the Knave of Hearts was wrong. We will show that this is logically impossible. Since the argument is lengthy, it will help to number the steps.

1. Suppose Bill the Lizard is right. Then his statement is true, meaning that either the March Hare or the Dormouse is right (or possibly both).
2. So, suppose the March Hare is right. Then the Cook must be right, because the March Hare said that the Cook and the Cheshire Cat are both right.
3. In contrast, suppose the Dormouse is right. This implies directly that the Cook is right, since this is part of what the Dormouse said. It follows that if Bill the Lizard is right, then so is the Cook.

4. Now, if the March Hare is right, then so is the Cheshire Cat. If the Dormouse is right, then so is the Caterpillar. This follows immediately from what they said. Since we know that at least one of the March Hare or the Dormouse is right, we also know that at least one of the Cheshire Cat or the Caterpillar is right.

5. In other words, either the Cheshire Cat or the Caterpillar is right. Since this is precisely what the Hatter said, we know that he is right as well.

6. This means the Cook and the Hatter are both right, which is precisely what the Knave of Hearts said. We have now shown that if Bill the Lizard is right, then so is the Knave of Hearts.

This shows that the White Rabbit was wrong in what he said. Given Alice's statement that the White Rabbit and the Duchess are either both right or both wrong, we conclude that the Duchess was wrong when she said that the Gryphon never stole the tarts. Therefore, the Gryphon *did* steal the tarts.

CHAPTER 8

From Aristotle to Russell

In this chapter, the plot of our story changes dramatically. A major part of Smullyan's output in recreational logic involved the empuzzlement of central principles in mathematical logic. I would like to tell you about that output, but doing so requires first that we discuss what mathematical logic actually is. This will be the work of several chapters.

You see, Aristotle worked out the basic principles of his logical system in the long-ago of ancient history, roughly the 300s BCE. The principles of propositional logic were worked out not long after that, by Stoic logicians working around 3 BCE (Bonevac and Dever 2012). For the next 19 centuries or so, the professional study of logic consisted largely of footnotes to this body of work.

Mostly absent from this history was any thought that mathematicians ought to be taking a special interest in logic. Certainly, mathematicians *used* logic in their daily activities, but it was never part of the zeitgeist that they should take a special interest in studying logic for its own sake. Also absent from this history was the notion that mathematical methods and symbolism could be useful to logicians. Logic textbooks through the ages looked like books meant to be read, as opposed to deciphered, despite their use of letters to abbreviate classes or propositions.

It was not until the nineteenth century that the walls separating mathematics from logic began to crumble. The history connecting Aristotle to the mathematical logicians of the nineteenth and early twentieth centuries is a long and winding one, and we shall recount a small portion of it here.

We will frame the discussion around the following statement, from philosopher Hilary Putnam:

[W]e first notice that logic, like every other science, undergoes changes—sometimes rapid changes. In different centuries logicians have had very different ideas concerning the scope of their subject, the methods proper to it, etc. Today the scope of logic is defined much more broadly than it ever was in the past, so that logic as some logicians conceive it, comes to include all of pure mathematics. (Putnam 1971, 1)

Our intent is to understand the broad strokes of the history of logic by considering what logicians of different eras chose to write about.

8.1 *Aristotle's* Organon

That portion of Aristotle's work that is today classified as "formal logic" is found primarily in the *Prior Analytics*. In Part II, we suggested that the minutiae of categorical syllogisms were perhaps not as interesting or important as Aristotle's attention to them suggested. You might even have wondered why someone as sharp as Aristotle evidently believed them to be worthy of such devotion. The answer comes from considering his logical works as a whole.

Aristotle did not write books in the modern sense. Rather, his writings were really more in the form of lecture notes intended for his students. After his death, various of his followers organized the notes into chapters and the chapters into books, until the notes assumed the form in which we know them today. As previously noted, six of those works were subsequently grouped together under the title *The Organon*, meaning "The Tool." It is the *Organon* as a whole that is, from a modern viewpoint, taken to represent Aristotle's work in logic. Surveying the *Organon* as a whole, we find a comprehensive treatise on human reasoning, as opposed to a narrow work of formal logic (Kneale and Kneale 1962, 23–25).

Aristotle took as his starting point that the physical world presents us with an astonishing number of objects and that these objects relate to one another in complex ways. A human reasoner desiring to think clearly about her world must somehow impose conceptual order on all of this messiness and confusion. This imposition of order takes place through various acts of the intellect.

The first such act is to apprehend the various objects of our world and to make distinctions among them. For example, we notice that some objects are living and some are not. Among those that are living, some are animals and some are not. Among the animals, some are vertebrates and some are not. It is this process of delineation that constitutes humanity's first attempt to impose order on the world. By drawing ever more careful distinctions, we eventually arrive at correct definitions for the objects we apprehend. In the language of Aristotelian metaphysics, objects have essences, these essences are as much a fact about their natures as are their chemical compositions, and a correct definition is one that accurately captures that essence.

The first work of the *Organon*, known as the *Categories*, is devoted to this first act of the intellect. Just as modern biologists seek the most general categories into which life should be classified (such as animal, plant, or fungus), so, too, did Aristotle seek to find the most general categories into which the objects of our apprehension ought to be arranged. He ultimately devised a list of ten such categories, but we will eschew presenting them here on the grounds that too much explanation would be required just to make the list comprehensible.

The second act of the intellect occurs when we formulate propositions asserting relationships among the objects. For example, we might say, "All cats are animals" or "No cats are fish." In this way, we assert relationships that exist among the objects of our apprehension, and by understanding the proper definitions of these objects, we can make judgments regarding the truth or falsity of the assertions.

The second work of the *Organon*, known as *On Interpretation*, is therefore devoted to the study of categorical propositions. It is here that Aristotle first formulates his distinctions regarding the quantity of a proposition (universal or particular) and the quality of a proposition (affirmative or negative). It is also here that Aristotle considers the operations of conversion, obversion, and contraposition of propositions, and presents his square of opposition.

Finally, we come to the third act of the intellect, which is reasoning. This is where we consider groups of propositions and draw conclusions as to what follows from them. The remaining books of the *Organon* are all, in some way, devoted to this topic. In the *Prior Analytics*, Aristotle develops his theory of the categorical syllogism. In so doing, he defined abstract argument forms that he viewed as a guide for all correct reasoning. This was followed by the *Posterior Analytics*, in which much of the machinery developed in the first three books is applied to questions of scientific reasoning. In the fifth work, the *Topics*, Aristotle considered questions regarding inductive reasoning, among other things. The sixth and final work, *On Sophistical Refutations*, addresses common logical fallacies.

If modern opinion of the subject were dictated entirely by the contents of the *Organon*, we would think that logic is a tool for organizing our apprehensions of the natural world. However, logic is more like a Swiss army knife than a single implement. One small aspect of logic is the enumeration of valid argument forms, but this is hardly the whole subject. In Aristotle's view, we are not studying logic to contemplate timeless abstractions valid in any possible world, whatever that means. Instead, our sole interest is to understand the world as we find it. This world presents us with objects, those objects have properties, and through various acts of the intellect, we make distinctions among those objects and properties by which we comprehend objective reality.

All human reasoning, in Aristotle's view, could be reduced to these three acts of the intellect, and this was why he placed so much emphasis on categorical syllogisms.

8.2 Medieval Logic

In the later Hellenistic period, a group of philosophers who came to be known as the Stoics produced numerous works on logic, advancing the subject beyond what Aristotle had accomplished. In particular, they expanded on Aristotle's narrow understanding of propositions as being exclusively categorical and developed theories of conjunctions, disjunctions, negations, and conditionals. It is from this work that propositional logic was born. In Stoic logic, the fundamental objects were propositions, and this distinguished it from Aristotelian logic, which was built on terms (O'Toole and Jennings 2004).

In the centuries following the decline of Greek civilization, little attention was paid to logic in Europe and the Mediterranean. The center of logical attention turned to India, China, and especially the Middle East. Arabic translations of Aristotle's works were produced, which is the only reason those works have survived to the present.

During the medieval period, generally taken to be roughly the years between the fifth and fifteenth centuries, we see a resurgence of interest in logic. The renewed interest in logic in Europe coincides with the emphasis by the Catholic Church on philosophical investigation, largely with an eye to supporting the claims of the religion (Kneale and Kneale 1962, 224–246).

Indicative of the growth of intellectual activity during this time is the production of works meant to organize and classify the various lines of scholarly inquiry that were pursued. One such work, written by Robert Kilwardby in the mid-thirteenth century, was entitled *De Ortu Scientiarum* (*The Rising Sciences*). It featured the following account of the history of logic:

> Since in connection with philosophical matters there were many contrary opinions and thus many errors (because contraries are not true at the same time regarding the same thing), thoughtful people saw that this stemmed from a lack of training in reasoning, and that there could be no certainty in knowledge without training in reasoning. And so they studied the process

of reasoning in order to reduce it to an art, ... and it is the science of the method of reasoning on all [subject] matters. (Kilwardby 1988, 265)

Work of seminal importance was produced during this period. Especially notable are the writings of Peter Abelard in the early twelfth century. In several works, he anticipated the development of what is now called classical logic by developing a truth-functional approach to propositional logic. He also developed a formal theory of logical consequence that helped unify the syllogistic logic of Aristotle with the argument forms involving hypothetical statements that had been developed by other writers. Many of the topics addressed in Abelard's logical works are still discussed in the logic textbooks of the present (Wilks 2008).

That said, the novelty of the logics produced in this era is largely found in their minute analysis of subtle questions in the philosophy of language. An example of an area of investigation that is distinctive to the medieval logicians is their development of the theory of supposition. As used by the writers of the time, "supposition" was essentially a synonym for the modern term "reference." In speaking of the "theory of supposition," we have in mind a classification of the various ways in which a written or spoken term might refer to some physical object or abstract concept. For example, sometimes a term simply refers to itself, as when I say: "'Cat' has three letters." Sometimes a term refers directly to a specific physical object, as when I say: "My cat is orange." And sometimes a term refers to a universal concept, as when I say: "A cat is a mammal." Respectively, these three examples are said to be instances of material, personal, and simple supposition. Medieval logicians found many other such distinctions to fuss over.

Let me be blunt, however, that the previous paragraph was a far clearer explanation of supposition than anything you will find in the literature of the time. The production of lucid and compelling writing was not on the agenda of any medieval logician who addressed this subject. Instead, they produced prose that was so turgid and murky that modern philosophers often struggle to explain even what they were saying, much less to explain why anyone thought it was important.

The tendency of medieval logicians to obsess over esoterica led to a backlash during the Renaissance. John Milton, best known as the author of *Paradise Lost*, spoke for many when he wrote, in a 1644 essay called "Of Education":

[Universities] present their young unmatriculated Novices at first comming with the most intellective abstractions of Logick and Metaphysicks: So that they having but newly left those Grammatick flats and shallows where they stuck unreasonably to learn a few words with lamentable construction, are now on the sudden transported under another climate to be tost and turmoil'd with their unballasted wits in fadomless and unquiet deeps of controversie, do for the most part grow into hatred and contempt of Learning, mockt and deluded all this while with ragged Notions and Babblements, while they expected worthy and delightful knowledge. (Milton 1644)

The old English spellings make the sarcasm all the more biting.

John Locke, in his *Essay Concerning Human Understanding*, first published in 1689, saw something sinister lurking behind the logician's obsession with language:

To this abuse, and the mischiefs of confounding the Signification of Words, Logick, and the liberal Sciences, as they have been handled in the Schools, have given Reputation; and the admired Art of Disputing, hath added much to the natural imperfection of Languages, whilst it has been made use of, and fitted, to perplex the signification of Words, more than to discover the Knowledge and Truth of Things: And he that will look into that sort of learned Writings, will find the Words there much more obscure, uncertain, and undetermined in their Meaning, than they are in ordinary Conversation. (Locke 1979, 493–494)

In 1662, an especially important logic textbook was produced by Antoine Arnauld and Pierre Nicole, though the book was initially published anonymously. It was called *Logic, or the Art of Thinking*. Arnauld and Nicole were associated with the Port-Royal Abbey in France, and the book has been known as the *Port-Royal Logic* almost since its publication. Over the next two centuries, it came to be seen as the most important textbook on logic since Aristotle, and it is often credited with reviving interest in the subject.

Like Milton and Locke, Arnauld and Nicole believed that logic had lost its way, and they were very eloquent in expressing their disdain. In response to the notion that logic as understood at that time provided a reliable route to proper reasoning, they write:

This is what the philosophers have specially undertaken to accomplish, and in relation to which they have made such magnificent promises. If we may believe them, they will furnish us, in that part which is devoted to this purpose, and which they call *logic*, with a light capable of dispelling all the darkness of the mind; they correct all the errors of our thoughts; and they give us rules so sure that they conduct us infallibly to the truth—so necessary, that without them it is impossible to know anything with complete certainty. These are the praises which they have themselves bestowed on their precepts. But if we consider what experience shows us of the use which these philosophers make of them, both in logic and in other parts of philosophy, we shall have good grounds to suspect the truth of their promises. (Arnauld and Nicole 1851, 6–7)

Having often expressed my dismay at the miserable writing so common of logic textbooks, the passion of the *Port-Royal Logic* comes as something of a relief. Also a relief is their belief that the study of logic has practical benefits, in that it helps people to see through the relentless nonsense put forth by the ignorant and the dishonest:

> Hence it is, that there are no absurdities too groundless to find supporters. Whoever determines to deceive the world, may be sure of finding people who are willing enough to be deceived; and the most absurd follies find minds to which they are adapted. After seeing what a number are infatuated with the follies of judicial astrology, and that even grave persons treat this subject seriously, we need not be surprised at anything anymore. (Arnauld and Nicole 1851, 2–3)

There follow several sentences of impressive sarcasm hurled in the direction of astrology. If only more logicians saw fit to write like this!

The *Port-Royal Logic* is composed of four major divisions. The titles of the first three will seem familiar:

- Containing Reflections on Ideas, or on the First Operation of the Mind, Which is Called Conceiving.
- Containing the Reflections Which Men Have Made on Their Judgments.
- Of Reasoning.

These sections contain the usual discussions regarding the signification of words, Aristotle's categories, the nature of propositions, and the figures and moods of the syllogism. These discussions go on for hundreds of pages, but the most important novelty to be found therein is probably their general attitude toward this material:

> There are other things which we deem sufficiently profitless; such as the categories and the laws, but which, as they were short, easy, and common, we did not think it right to omit, forewarning the reader, however, what judgment to form of them, in order that he might not suppose them to be more useful than they are. (Arnauld and Nicole 1851, 8)

There is more novelty to be found in the book's fourth section, titled "Of Method." By this they mean the method for applying the more abstract considerations of the book's first three parts to various practical problems. By including this material, Arnauld and Nicole showed an admirable attention to real-world concerns that was absent from most then-recent work in logic. They focus especially on certain alleged pecca-dilloes committed by geometers, as well as on a defense of religion and miracles from the skeptical attacks common to their age.

Arnauld and Nicole brought a new level of passion and relevance to logical analysis, and they were more clearheaded than their contemporaries about what was important and what was not. For all of that, however, logic after the *Port-Royal Logic* was scarcely different from logic before. It remained a highly abstruse amalgam of topics from the philosophy of language, metaphysics, and epistemology.

8.3   *Mill's* A System of Logic

In 1843, British philosopher John Stuart Mill published a massive work typically referred to simply as *A System of Logic*. The full title was this: *A System of Logic, Ratiocinative and Inductive: Being a Connected View of the Principles of Evidence and the Methods of Scientific Investigation*. In this context, "ratiocinative" is synonymous with "deductive." He described the goal of his work as follows:

> Our object, then, will be to attempt a correct analysis of the intellectual process called Reasoning or Inference, and of such other mental operations as are intended to facilitate this: as well as on the foundation of this analysis, and *pari passu* [side by side] with it, to bring together or frame a set of rules or canons for testing the sufficiency of any given evidence to prove any given proposition. (Mill 1974, 12)

To flesh out what this means, Mill elsewhere explains that among the operations he will consider are those surrounding the use of language, such as naming objects, defining them, and classifying them. He writes: "Language is evidently... one of the principal instruments or helps of thought; and any imperfection in the instrument, or in the mode of employing it... [will] destroy all ground of confidence in the result (Mill 1974, 19). Given this understanding of his project, it is unsurprising that Mill, following the example set by so many other logicians, opens with a great many pages devoted to the understanding of terms, propositions, and syllogisms.

Mill had many insightful things to say on these matters, but it is when he mentions the testing of evidence that we come to something largely new. More so than previous writers, Mill emphasized the importance of inductive reasoning. In his view, Aristotle's syllogistic encompassed the whole of deductive reasoning, and while it could be useful as a device for organizing our thoughts, it could never hope to provide new knowledge. This is because the conclusion of a syllogism provides nothing that was not already implicit in the premises. Instead, inductive reasoning, in which we move from observations of particulars to general conclusions, allows us to learn something new. Mill writes:

> The conclusion in an induction embraces more than is contained in the premises. The principle or law collected from particular instances, the general proposition in which we embody the result of our experience, covers a much larger extent of ground than the individual experiments which form its basis. A principle ascertained by experience is more than a mere summing up of what has been examined. (Mill 1974, 163)

That someone like Mill would come along at this time to elevate the status of inductive reasoning is unsurprising when we consider the scholarly climate in Europe in the mid-nineteenth century. The dominance of Aristotelian notions in logic and metaphysics had been enforced in large measure by the Catholic Church's embrace of these doctrines. Starting in the Renaissance, the Church gradually lost much of its political power, leading to a newfound freedom to challenge these ideas.

In philosophy, this freedom expressed itself in part through a disdain for Aristotle's syllogistic as a model for human reasoning. To be clear, the objection was not that philosophers had suddenly come to reject the validity of argument forms like Barbara or Celarent (recall Section 3.4). Instead, it was that philosophers had come to reject the metaphysical assumptions underlying Aristotle's approach to logic, and this rejection made syllogistic reasoning come to seem trivial and irrelevant. Philosopher Fred Wilson describes the situation:

> This metaphysics of forms and substances was subject to critique in the ancient world, but the sceptical tradition largely disappeared during the long period during the middle ages and beyond when philosophy was subordinated to theology. In the early modern period, the sceptical critique was revived by Montaigne, and developed more systematically by a series of British thinkers, Bacon, Hobbes, Locke, Berkeley, and Hume. (Wilson 2008, 235)

A second development influencing Mill was the rapid pace of scientific advancement during his life. The Catholic philosophers who dominated the universities of the medieval period were primarily interested in theology and religion. Systematic investigation of nature was important only to the extent that it served religious ends. Historian David Lindberg describes the Church's attitude as follows:

In Augustine's influential view, then, knowledge of the things of this world is not a legitimate end in itself, but as a means to other ends it is indispensable. The classical sciences must accept a subordinate position as the handmaiden of theology—the temporal serving the eternal. . . . This attitude toward scientific knowledge came to prevail throughout the Middle Ages and survived well into the modern period. (Lindberg 2009, 16)

In such an environment, it is understandable that philosophical analysis generally, and developments of logic in particular, would turn toward highly abstract problems of little practical significance.

With major scientific advances arriving quickly at the same general time that old philosophical certainties were seen as shaky and unhelpful, it is hardly surprising that someone would think to pay careful attention to the sort of reasoning underlying science. This reasoning was uniformly inductive, with the deductive reasoning of the old logic largely irrelevant.

Mill's tome was the last serious work of logic to be carried out to a significant degree within the Aristotelian tradition. Textbooks presenting syllogistic logic to students continued to be written well into the twentieth century, but the frontiers of the subject soon came to be dominated by different methods altogether.

8.4   Boole and Venn

In 1854, George Boole published *An Investigation into the Laws of Thought, on Which are Founded the Mathematical Theories of Logic and Probabilities*. Boole's was the first major treatise on logic that would appear, to someone flipping casually through its pages, to be a work of mathematics.

The basic idea of developing a logical calculus (by which is meant a system for reducing logical analysis to a process of mechanical symbol manipulation) was pioneered by Gottfried Leibniz in the seventeenth century. Boole does not seem to have been aware of Leibniz's work when writing his own book. The general consensus among logicians of the nineteenth century was that Boole's work represented a major development of the ideas that Leibniz introduced (Peckhaus 2018).

In an introductory chapter, Boole characterizes the history of logic as follows:

In its ancient and scholastic form, indeed, the subject of Logic stands almost exclusively associated with the great name of Aristotle. As it was presented to ancient Greece in the partly technical, partly metaphysical disquisitions of the Organon, such, with scarcely any essential change, it has continued to the present day. The stream of original inquiry has rather been directed towards questions of general philosophy, which, though they have arisen among the dispute of the logicians, have outgrown their origin, and given to the successive ages of speculation their peculiar bent and character. (Boole 1951, 1–2)

Like so many before him, he characterized logic as the branch of inquiry seeking to formulate principles of correct reasoning, or "the laws of thought," as he put it. In his view, "the ultimate laws of logic are mathematical in their form" (Boole 1951, 11). The remainder of the book was then devoted to showing how standard algebraic manipulations could be given a logical interpretation.

Here is a taste of what Boole had in mind: Let us think of x and y as representing two classes to which an object might belong. We might then interpret xy to represent the class of things belonging to both x and y. Likewise, we could think of $x + y$ as representing the class of things that belong either to x or to y. (For technical reasons, it was necessary to assume that x and y were disjoint when using this notation.)

For example, suppose x represents the class of cats and y represents the class of dogs. Let z represent the class of things that are black. If we treat x, y, and z as standard algebraic variables, then we can write the equation:

$$z(x + y) = zx + zy.$$

Algebraically, this is an instance of the distributive property. But given our logical understanding of the terms, we can construe the left side as representing the class of things that are both black and either a cat or a dog. On the right side, we have the class of things that are either black cats or black dogs. This makes sense. If you are a black thing that is either a cat or a dog, then you are either a black cat or a black dog.

What happens if we multiply a class by itself? Plainly, we get the original class back again. That is, if we ask, "What is the class of things that is in the class x and also in the class x?" then we must reply that we have done nothing more than describe x itself. Using standard algebraic notation, this observation can be represented by the equation $x^2 = x$.

The only numbers that satisfy this equation are 1 and 0, and therefore these are the only numbers that play a role in Boolean logic. These numbers have natural logical interpretations. We can think of 0 as representing the empty class, which contains nothing, while 1 can be seen as representing the entire universe of discourse. So, if x represents a class, then we can think of $1 - x$ as representing the collection of things that are not in x.

Let us go back to the equation $x^2 = x$. Algebra tells us that this is equivalent to $x - x^2 = 0$, which in turn is equivalent to $x(1 - x) = 0$. In words, this equation says that the class of things that is simultaneously in x and not in x is the empty class. Just like that, we have recovered the law of noncontradiction.

Now suppose that $f(x)$ denotes an algebraic expression that involves x and possibly some other terms. We can think of $f(x)$ as a function in which x may only take on the values 0 or 1. Then we can write the equation:

$$f(x) = f(1)x + f(0)(1 - x).$$

To see that this is true, note that both sides give the same result when we set $x = 1$ and also give the same result when we set $x = 0$. It follows that if we ever have an equation of the form $f(x) = 0$, then we can rewrite the equation in the form $f(1)x + f(0)(1 - x) = 0$. This can be solved for x to obtain

$$x = \frac{f(0)}{f(0) - f(1)}.$$

This observation has an interesting consequence. For we now see that

$$1 - x = 1 - \frac{f(0)}{f(0) - f(1)}$$

$$= \frac{f(0) - f(1)}{f(0) - f(1)} - \frac{f(0)}{f(0) - f(1)} = \frac{-f(1)}{f(0) - f(1)}.$$

The law of noncontradiction now implies that

$$x(1 - x) = \frac{-f(0)f(1)}{(f(0) - f(1))^2} = 0,$$

which in turn implies that

$$f(0)f(1) = 0.$$

Do you see what we just accomplished? Starting from an equation of the form $f(x) = 0$, we obtained a second equation in which x plays no part. In logical terms, we eliminated one variable and derived an expression that only involves the others. Accomplishing this was precisely the "elimination problem" considered by Lewis Carroll (as discussed in Chapter 5).

Let us apply this result to the problem of analyzing categorical syllogisms. Suppose that A, B, and x represent three terms. What can we conclude from the premises "All As are x" and "All xs are B"?

We first need to translate these categorical propositions into our algebraic notation. To assert that all As are x is to say that it never happens that being in A is found in conjunction with not being in x. This can be expressed by the equation $A(1 - x) = 0$. Likewise, our second premise becomes $x(1 - B) = 0$. We can now reason as follows: We are given

$$A(1 - x) = 0$$
$$x(1 - B) = 0,$$

from which it follows by simple addition that

$$A(1 - x) + x(1 - B) = 0.$$

Now define

$$f(x) = A(1 - x) + x(1 - B).$$

We can check that $f(0) = A$ and $f(1) = 1 - B$. It follows that

$$0 = f(0)f(1) = A(1 - B),$$

which asserts that "All As are B." We have just rediscovered Barbara (see Section 3.4).

That was actually kind of fun! Let us try Celarent, with the premises "All As are x" and "No xs are B." The equations that express these premises are $A(1 - x) = 0$ and $Bx = 0$. We can combine the premises to get

$$A(1 - x) + Bx = 0.$$

Setting $f(x) = A(1 - x) + Bx$ and once more using the equation $0 = f(0)f(1)$, we get simply that $AB = 0$. In words, "No As are B," which is the correct answer.

In this way, the entirety of Aristotle's syllogistic can be recovered simply by manipulating a few equations. Boole's method was more powerful than this, however, and could be used to analyze arguments that were more complex than those pondered in traditional logic. Moreover, Boole's methods proved fruitful for analyzing arguments drawn from the theory of probabilities. In short, Boole's work showed that Aristotle's syllogistic was only one small part of logical analysis as a whole, and Boole produced a fundamentally new approach to the problems he considered.

Of course, it should go without saying that there are philosophical difficulties in some of Boole's arguments. For example, it might have bothered you that a few paragraphs ago, I was rather cavalier about dividing my logical symbols, even though the operation of division has no obvious logical interpretation. It bothered Boole, too, and he made some attempts to address the question. Other philosophers then challenged those attempts, but this minutiae is not relevant for the progress of our story.

Boole's work proved controversial among logicians of his time. There were some who argued, contemptuously, that the reliance not just on symbols, but on algebraic manipulations, consigned Boole's work to the fever swamps of mathematics, as opposed to the high tower of logic.

In 1881, John Venn published a book called *Symbolic Logic*. The book is a lengthy explication and development of Boole's ideas, best remembered for containing a thorough discussion of the sorts of logic diagrams that bear Venn's name to this day. However, I wanted to mention his book mostly for the way it opened:

> There is so certain to be some prejudice on the part of those logicians who may without offence be designated as anti-mathematical, against any work professing to be a work on Logic, in which free use is made of the symbols + and − and × and ÷ (I might almost say, in which x and y occur in place of the customary X and Y) that some words of preliminary explanation and justification seem fairly called for. Such persons will without much hesitation pronounce a work which makes large use of symbols of this description to be mathematical and not logical. (Venn 1881, ix)

Later he elaborates:

> Hence it comes about that when people meet with such an expression as $x + y$ in a work which is professedly logical, they are apt to say to themselves, 'This is a mathematical symbol: the author is diverging into mathematics, or at least borrowing something from that science.' If they were to explain themselves they would probably maintain that the sign (+) stands for addition in mathematics, and therefore should not be resorted to when we have no idea of adding things together. (Venn 1881, x)

Venn goes on to explain, with great eloquence, the foolishness of this argument. Even within mathematics, he argues, the same symbols are used for very different purposes, simply because the same symbolical calculus can be interpreted in many ways. In applying symbolic methods to logic, what matters are the ideas being expressed, and not the precise symbols used to express them.

As one of my undergraduate professors put it, we care about notions, not notations!

There is something adorable in the idea that as recently as 1881, Venn had to explain these simple points to the very brilliant people who undertook serious work in logic. Modern scholarly journals in logic are today indistinguishable from mathematics journals. Indeed, an article on logic that did not include copious helpings of mathematical notation would probably be deemed unserious for that reason alone.

I have a second reason for including the books by Boole and Venn in this survey: They represented another advance over the logical works of the time. That advance is this: Simply as works of literature, they are far superior to what most logicians, then or now, were capable of writing. If you have a taste for logical analysis, and if you appreciate clear, lucid prose, then you will find the works of Boole and Venn to be surprisingly enjoyable reading.

8.5 *Russell's* The Principles of Mathematics

In 1903, Bertrand Russell published *The Principles of Mathematics*. Early in the book, he wrote the following:

> [M]athematics includes not only Arithmetic and Analysis, but also Geom-etry, Euclidean and non-Euclidean, rational Dynamics, and an indefinite number of other studies still unborn or in their infancy. The fact that all Mathematics is Symbolic Logic is one of the greatest discoveries of our age; and when this fact has been established, the remainder of the principles of mathematics consists in the analysis of Symbolic Logic itself (Russell 1903, 5).

I previously mentioned the ferment in the philosophical community during the eighteenth and nineteenth centuries, and how it led to new ideas regarding the proper focus of logical investigations. During roughly the same time, mathematicians were forced to confront some crises of their own.

For many centuries, Euclid's system of geometry was for mathemati-cians what Aristotelian logic was for philosophers. It was seen as a correct description of how the world really was. To the extent that non-Euclidean geometry was even coherent, it could only be a silly diversion from more important concerns. This consensus was challenged on two fronts in the nineteenth century: Some of Euclid's reasoning was found to be faulty, and useful models of non-Euclidean geometries were developed by several researchers.

Likewise, the field of analysis (calculus) started to produce some dis-concerting results. Noteworthy in this regard was the discovery by Karl Weierstrass of a function that was continuous at every point, but not differentiable at any point. That such a thing was possible was surpris-ing to the researchers of the time. It showed that the vague, intuitive notions regarding the nature of functions, on which mathematicians so often relied, were not sufficient (George and Van Evra 2002).

These sorts of considerations led mathematicians to fear they had paid insufficient attention to the foundations of their subject. Theorems and proofs ultimately rest on a foundation of undefined terms and unproved assumptions. If the foundation itself contained contradictions, then everything built on it was nonsense, even if very clever nonsense. Thus, the idea of finding the firmest possible foundation for mathematics was very attractive, and this was the project Russell attempted in *The Principles of Mathematics*.

The idea then developed of reducing mathematics to logic. By this is meant that a formal language was sought in which all the truths of mathematics could be expressed, and, on the basis of a few basic logical principles, proved. The systems of propositional logic available at that time could readily handle the standard logical constructions of natural language, such as conjunctions, disjunctions, conditional statements, and negations. This was powerful stuff, but it lacked certain resources that were needed for mathematics.

For example, there was no mechanism for expressing relations among terms. An argument such as "Joe is Bill's father; therefore Bill is Joe's son" could not be formalized in the logic of the time. A favorite example of Russell's that was used to make the same point is: "A horse is an animal; therefore the head of a horse is the head of an animal." A second problem was that there was no mechanism for handling quantified statements. Constructions such as "For all real numbers x, there exists a real number y, such that P," where P is some unquantified proposition, could not then be expressed formally.

It was these two deficiencies of propositional logic that needed to be corrected before the reduction of mathematics to logic could take place. A major step in that direction was the publication of Gottlob Frege's 1873 book the *Begriffsschrift*, whose full title is very long and hard to translate. But it was in this book that Frege attempted for the first time to develop theories of relations and quantification that would be sufficient for expressing all of mathematics. Russell brought Frege's work to a much wider audience than it had previously enjoyed, and also developed its ideas far beyond what Frege had considered.

The early chapters of *The Principles of Mathematics* develop a formal system in which mathematics can be modeled. Russell devotes considerable space to discussing terms and propositions, but there are only a few brief mentions of the syllogism, and these were mostly for the sake of examples or to note that the Aristotelian system was insufficient for mathematical purposes. From there, the book moved on to discuss a bewildering array of topics in mathematics and physics, all with the intent of showing they were ably modeled in his symbolic language.

Russell later came to see *The Principles of Mathematics* as a rough and immature work. Together with Alfred North Whitehead, he would later put forth a more sophisticated version of this material in a three-volume work called *Principia Mathematica*. The formal system introduced in the *Principia* is today referred to as "classical logic."

It was with the publication of the *The Principles of Mathematics*, however, that mathematical logic became the dominant paradigm for logicians. After Frege and Russell, Aristotelian logic was suddenly of interest only to historians of the subject.

8.6 Notes and Further Reading

As an academic discipline, the history of logic has reached a high state of development. The *Handbook for the History of Logic* comes to 11 large volumes and looks like a set of encyclopedias (for readers old enough to understand that reference). It is an incredibly useful source for high-level discussions on any aspect of the history of logic. Also useful is the book by Kneale and Kneale (1962). These references make for challenging reading. For a more gentle discussion of the history of logic, try Shenefelt and White (2013).

CHAPTER 9

Formal Systems in Life and Math

The brief history of logic in Chapter 8 culminated with Russell and others striving to reduce all of mathematics to a small number of axioms from which everything else could be derived. That this is flatly impossible was shown by Kurt Gödel in 1931. The theorems showing this are nowadays referred to as his first and second incompleteness theorems.

Raymond Smullyan was a great authority on the subtleties of Gödel's theorems, and he devoted much of his popular writings to presenting the underlying ideas in the form of logic puzzles. Some of these puzzles employed his usual device of knights and knaves, while others involved puzzles based on the mathematical theory of computation.

This chapter and the next one are devoted to Smullyan's puzzles in this area. The present chapter is about understanding what formal systems are and why mathematicians find them interesting. Chapter 10 switches the focus to Gödel's theorems and to the puzzles Smullyan devised for explaining them.

9.1   What Is a Formal System?

Gödel's theorems are about properties of "formal systems." Let us consider some nonmathematical examples of what this means.

The rules of chess define certain abstract objects. There is something called "the chessboard," which, we are told, is divided into 64 small squares. There are pieces called "kings," "bishops," and "knights," among others, and rules are laid down governing their legal movements around the board. Abstract concepts such as "check" and "checkmate" also play a role in the game.

Chess is an example of a formal system. The objects and concepts laid down by the rules exist only in the context of the game. If someone asked, "What are bishops and knights *really*?" or asked you to show them a chess bishop or knight in real life, you would have to reply that the question does not make sense. Bishops and knights have no existence outside a game of chess, and everything there is to know about them is contained in the rules.

In everyday life, when we speak of a "formal occasion," or when we suggest that "formal attire" is appropriate, we are referring to certain conventions that are understood within a society. In the United States, for example, any man told to dress formally would just know from experience to wear a business suit or tuxedo. He would wear stylish loafers on his feet, as opposed to running shoes. He would do these things not because it is an objective fact about the world that formal attire consists of business suits and dress loafers, but because he is aware of the social conventions that govern such events. Contrast this with the physical systems studied by scientists. It is not a social convention that a water molecule is two parts hydrogen to one part oxygen.

The rules of chess are more like social conventions than they are like scientific facts. Bishops and knights move as they do not because those movements are objectively correct, but only because everyone just agrees on how the game is to be played. This is why it is appropriate to refer to the rules of chess as a "formal system."

Once the rules are laid down and understood, we might discover they entail certain conclusions that we did not anticipate while formulating them. For example, it turns out that the rules of chess entail that a king, bishop, and knight can force checkmate against a lone king, but that a king and two knights cannot. The rules of chess are especially rich in such surprising consequences, which is why people continue to study them so many centuries after they were first laid down.

Another familiar context in which something like a formal system arises is a fictional story. It is commonplace for people to discuss fictional characters as though they had a real existence. For example, they might have a passionate argument about what James Bond would or would not do in a given situation, despite knowing that he has no existence outside the fictional world created by Ian Fleming (as enhanced by the films, of course).

Questions about fictional characters are unintelligible if construed as addressing actual people in the physical world, but they make sense when thought to be addressing the formal world created by the character's author. Sometimes the author might provide enough information to answer the question directly—Did Sherlock Holmes smoke a pipe? Did Captain Kirk have a half-Vulcan first officer? Other times we must draw

such inferences as we can from the information the author provides—
Does Harry Potter like to eat spaghetti? Regardless, it is understood that
our reasoning only makes sense in the context of a fictional reality whose
nature has only those arbitrary properties an author decided it should
have.

This analogy illuminates another aspect of formal systems. In a long-
running book or television series, it sometimes happens that the events
of later episodes contradict those of earlier episodes. Such inconsistency
is intolerable and must be resolved. In the context of a fictional series,
everyone understands that, really, the inconsistency is the result of nothing
more than later writers having forgotten what earlier writers had established. But it can still be fun to devise arguments for why the inconsistency
is only illusory, resolvable by careful investigation of the series' minute
details. However, in the abstract formal systems of interest to mathematicians, the specter of inconsistency is a constant menace. It is nothing to
joke about.

Moving from everyday life to abstract mathematics, in most cases, students encounter formal systems for the first time in a geometry class. It
is in such a class that proof becomes a major focus of discussion. When
someone tells you about the Pythagorean theorem, or tells you that the
area of a circle is πr^2, you do not just passively accept it. Instead, you
insist on being shown the reasoning behind the assertion. You want to
see the proof.

But even beginning students quickly realize that the demand for proof
must eventually terminate at certain statements that are just taken as
given. Complex theorems are proved by reference to simpler theorems,
which are proved from simpler theorems still. Eventually you reach assertions so simple that there is nothing simpler from which to prove them.
These too-simple-to-prove assertions are referred to as *axioms*.

It was the Greek mathematician Euclid who first attempted, in roughly
300 BCE, to devise a formal system sufficient for proving all theorems
then known about geometry. People had been studying geometry for several centuries before Euclid arrived, and they had produced a large body
of useful findings. Some portions of this work were justified empirically,
while others were the result of abstract reasoning. Euclid sought to organize this work into one coherent theory. To do that, he needed to identify
the minimal logical commitments that were needed to derive everything
that was then known (or, more accurately, believed).

Euclid began by providing definitions for fundamental terms, such as
"point," "line," and "right angle." He then formulated five axioms which,
he believed, were sufficient for deriving the entirety of geometry. Here are
those axioms:

(1) A straight line segment can be drawn joining any two points.
(2) Any straight line segment can be extended indefinitely in a straight line.
(3) Given any straight line segment, a circle can be drawn having the segment as radius and one endpoint as center.
(4) All right angles are congruent.
(5) Given a line, and a point not on the line, there is exactly one line through the point that is parallel to the given line.

Let us note two aspects of this list, before continuing.

First, what I have given as the fifth postulate was not actually what Euclid said but is instead a simpler statement logically equivalent to Euclid's version. Second, the fourth postulate might seem mysterious to a modern reader, since nowadays we are accustomed to defining a right angle to be one that measures 90 degrees. By that definition, it seems obvious that all right angles are congruent. Euclid's definition was different, however. In his usage, if two lines intersected in such a way that the resulting angles were equal, then those angles were said to be right. Now, if two lines *over here* intersect in a manner that makes the resulting angles congruent, and two other lines *over there* do likewise, there is nothing to guarantee that the angles from the first pair equal the angles from the second. Hence the fourth postulate.

Returning to our discussion, Euclid's formal system also involved certain "common notions." By this he meant statements such as "Things which equal the same thing are equal to each other." In other words, his common notions were certain principles of reasoning so fundamental that they are taken as given.

Now, Euclid's idea was that once you accept his definitions, axioms, and common notions, all the theorems of geometry follow as a matter of logic. There is certainly much to be challenged in that idea. Bertrand Russell was especially scathing, writing:

> His definitions do not always define, his axioms are not always indemonstrable, his demonstrations require many axioms of which he is quite unconscious. A valid proof retains its demonstrative force when no figure is drawn, but very many of Euclid's earlier proofs fail before this test.... [T]he value of his work as a masterpiece of logic has been very grossly exaggerated. (Russell 1902, 165)

Substantively, Russell is correct. For example, Euclid defined a *line* as "breadthless length," and a *straight line* as "one that lies evenly with its points." It is doubtful that someone not already aware of Euclid's intention would comprehend it based on these definitions. Russell was likewise clever about noting logical infelicities in Euclid's work. Still, it seems churlish not to recognize the revolutionary nature of what Euclid accomplished. Since he was the first to attempt a mathematical systematization of this magnitude, it is surely forgivable that he overlooked a few things.

At any rate, for our purposes, there are two things to stress about Euclid. The first is that the development of an axiomatic system for a branch of mathematics arises rather late in its development. Mathematicians do not simply write down random axioms in the hope that an interesting body of theorems will arise. Instead, the actual sequence is this: First, a body of theorems is developed through some combination of empirical experience and plausible (if not logically rigorous) argument. Later, once the theory has proven its worth, there is some value in organizing everything carefully and in discerning the minimal set of assumptions necessary for proving the desirable theorems.

The second point is that we usually have some concrete idea in mind before devising our abstractions. While reading Euclid's definitions, you can almost feel his exasperation. His desire for a completely formal system was sometimes at odds with his understanding that everyone was already familiar with the fundamental objects that geometers study. It is as though he wanted to grab his readers by the shoulders and say, "Come on! You know perfectly well what I mean by points and lines!" He had in mind specific instantiations of these concepts, inspired by the many basic shapes he saw in the world.

Though Euclid recognized the need for unproved assertions—axioms—at the base of his theory, he did not likewise appreciate the need for undefined terms. Nowadays we would simply leave terms like "point" and "line" undefined. Our chess analogy is apt here. We did not first define what bishops and knights were before explaining how they moved. They were simply defined entirely by what the rules say they can do in the game. Likewise, we can say, with regard to Euclid's first axiom, "I have no idea what points and lines are, but if you give me two points I guarantee there will be a line joining them, whatever any of that means."

Apropos of this last point, let us consider one more example of a formal system used by mathematicians.

We shall say that a *group* consists of a set S and a binary operation $*$ that satisfy the following properties:

(1) If a, b, and c are three elements of S, then $a * (b * c) = (a * b) * c$.
(2) There is an element e in S, such that $a * e = e * a = a$, for any element a of S.
(3) For every element a in S, there is an element b in S such that $a * b = b * a = e$.

Recall that a *binary operation* in mathematics is something like addition or multiplication of whole numbers. It takes two items from a set and combines them in some way to produce a third element of the set (not necessarily different from one of the original two).

If you have never seen the definition of a group before, it can be a lot to take in. For our present purposes, the usefulness of this example lies in how perfectly abstract it is. This definition is typically presented very early in a course in abstract algebra, well before the student has any preconceived notion of what the letters are supposed to represent. This is very different from Euclid's system of geometry, where we go into the subject understanding perfectly the objects that are under discussion.

One example of a group is the set of integers (positive, negative, and 0) with normal addition as the operation. The number 0 plays the role of e from axiom (2), since $x + 0 = 0 + x = x$ for any integer x. And if y is any integer, then the number $-y$ has the property that $y + (-y) = (-y) + y = 0$. A different example of a group is the set of rational numbers, other than 0, with the operation being normal multiplication. This time the number 1 plays the role of e (since multiplying a rational number by 1 leaves the number unchanged), and the rational numbers a/b and b/a have the relationship to each other described in axiom (3). (We had to omit the number 0 from this group, because including it would violate axiom (3). It is impossible to multiply 0 by something to obtain 1.)

The letters used in the definition do not have to be numbers at all. If we are doing linear algebra, then they might represent matrices. If we are doing calculus, the letters might represent continuous functions. If we are doing geometry, they might represent symmetries of a plane figure. There is almost no end to the number of interpretations we can give to the letters.

And for all of that, when we study abstract groups, we do not care at all about what the letters represent. We care only about the rules they follow.

We care only about their role in the game.

9.2 What Can Your Formal Language Say?

There is an irony in this treatment. Though I have been extolling the virtues of formal systems, my treatment of them has been decidedly informal. The problem is that I have mostly expressed the axioms in plain English which, like all natural languages, is beset by vagueness and imprecision at every turn.

If we *really* want to do this right, then we should work instead with a formal language. That is, we should state precisely what our undefined terms are, and then define the precise manner in which those undefined terms can be strung together to form grammatically correct statements, typically referred to as *well-formed formulas*, or *wffs* for short. Though it will be in the back of our minds that our well-formed formulas are meant to model some interesting aspect of natural language, the formal language itself will be treated as uninterpreted. When working with the formal language, regardless of our motivation for creating it, let us envision only arbitrary symbols manipulated according to arbitrary rules. It is only in this way that we can eliminate the problems of vagueness and ambiguity.

However, in solving the vagueness problem, we have created a new one. Specifically, we now have to think carefully about what notions our language is capable of expressing. What should our language be able to say?

For example, we have spent some time discussing propositional logic. In this system, we use capital letters to represent basic propositions. We then define connectives, such as $\vee$, $\wedge$, $\rightarrow$, and $\neg$, and explain how the truth of formulas expressed with these connectives relates to the truth of the basic propositions. This is a powerful language. It allows us to express many complex thoughts, and it brings clarity to the logical structure of those thoughts. In particular, it captures much of the logic that we employ when doing mathematics.

However, as mentioned in Chapter 8, there are also mathematical notions that cannot be expressed in this language. It has difficulty with quantification, for example. I cannot express a notion such as "There exists an odd perfect number" or "All prime numbers except for two are odd." The phrases "there exists" and "for all" are unintelligible in the formal language of propositional logic. Also problematic are propositions that express relations among variables. How do I say, "Five is greater than three but less than seven"? Propositional logic has never heard of "greater than" or "less than." The best it could do is to rephrase the sentence as "Five is greater than three and five is less than seven." Then we might denote "five is greater than three" as P and "five is less than seven" as Q. Our statement now becomes $P \wedge Q$. Plainly, something important from our original statement has been lost in this translation.

Since relations and quantification are unavoidable in mathematics, propositional logic is insufficient for mathematicians. We therefore enhance the language with further symbols. We define symbols such as $\exists$ and $\forall$, to capture notions of quantification. Of course, having defined the symbols, we also put forth rules for assessing the truth values of statements containing them. We also add some concept of predicate logic to our system, so that we can capture notions of relations among variables. The resulting system is typically referred to as *classical logic*. It provides a language sufficient for expressing everything mathematicians wish to say, at least in their professional lives.

9.3   Formalizations of Arithmetic

The question of precisely which ideas can, or cannot, be expressed in a given formal system is at the heart of Gödel's incompleteness theorems. In particular, he studied formal systems capable of expressing the fundamental ideas of arithmetic. What attributes would such a system need to have? What does it need to be able to say? What should be laid down as axiomatic?

Reminiscent of our discussion of Aristotelian logic in Chapter 3, you should not feel obligated to master the technical details of what follows. What matters is the principle that we can reduce what is essential about the arithmetic of natural numbers to a few concise axioms.

This is how we normally think of the natural numbers (assuming that we include 0 among them):

$$0, 1, 2, 3, 4, 5, 6, 7, \ldots.$$

Take a moment to describe what you see.

Perhaps the first thing we notice is that the numbers come in a definite sequence. Wherever we are in the middle of the sequence, we add 1 to get to the next, or subtract 1 to get to the previous number. This is different from some other familiar number systems, such as the rational numbers or the real numbers. In those systems, there is no well-defined next number after the one you are sitting on. There is an exception to our ability to add or subtract 1, however. The number 0 starts the whole sequence. There is no previous number to which we add 1 to get to 0.

This suggests that our formal system might begin like this: We will have the special symbol 0, which, of course, represents the number zero. We will also define a function $s(x)$, which we think of as taking any natural number to the next one in the sequence. So, $s(4) = 5$, and $s(17) = 18$, just to give two examples. Let us refer to this as the successor function. We can now write down a few axioms to define the sequence of natural numbers. (For convenience, I will continue to express axioms in plain English, rather than in a purely symbolic language. Also, let us take for granted that we know what the = sign means.)

- 0 is a natural number, but it is not the successor of any number.
- If $s(x) = s(y)$, then $x = y$.
- If x is not 0, then x is the successor of some number y.

Expressed more simply: The number 0 begins the sequence, and every number after that has both a unique successor and a unique predecessor.

We can now express some axioms that will define our basic notions of arithmetic. We will freely use the symbols + and × in the normal way.

Intuitively, we want to say that addition is closely related to the successor function, in the sense that $s(x) = x + 1$. Adding arbitrary numbers can then be viewed as applying the successor function multiple times. We also want to say that adding 0 does not change anything. These ideas can be captured with two further axioms:

- If x is any natural number, then $x + 0 = x$.
- If x and y are any two natural numbers, then $x + s(y) = s(x + y)$.

The second bullet point is really just saying that $x + (y + 1) = (x + y) + 1$.

Finally, once we have addition, we can also define multiplication. Not wishing to belabor things beyond decency, here are some axioms that do the job:

- If x is any natural number, then $x \times 0 = 0$.
- If x and y are any two natural numbers, then $x \times s(y) = (x \times y) + x$.

As with the axioms for addition, the second bullet point becomes clear when you realize that it is just expressing the distributive property, namely, that $x \times (y + 1) = (x \times y) + x$.

We now have axioms that tell us what the natural numbers are, and other axioms that tell us how addition and multiplication behave. Within this system we could now, in principle, prove statements like $2 + 3 = 5$, where we take for granted that $2 = s(s(0))$, that $3 = s(2)$, and so forth. However, there is one aspect of the logical structure of the natural numbers that has not yet been covered. It is typically referred to as the principle of induction.

Let us use the notation $\phi(x)$ to denote a statement that may or may not be true of any given natural number x. For example, we might construe $\phi(x)$ to be: "The number x is even." Then $\phi(x)$ will be true if $x = 0, 2, 4, \ldots$ and will be false if $x = 1, 3, 5, \ldots$. We can say that $\phi(x)$ is not a proposition, but that it becomes a proposition as soon as a specific number is substituted for x.

With that notation, we can now state the principle of induction:

- Suppose that $\phi(0)$ is true. Further suppose that if x is any natural number, then we know that the truth of $\phi(x)$ implies the truth of $\phi(x + 1)$. Then we know that $\phi(x)$ is true for any natural number.

This axiom is often analogized to knocking over dominoes. If you knock over the first one, and if you know that each time one domino falls the next in line also falls, then you know that all the dominoes will fall.

This collection of axioms is typically referred to as the system of Peano arithmetic, or **PA** for short, after the nineteenth-century Italian mathematician who first devised it. It seems to capture all of our intuitions about how the natural numbers behave.

Suppose, though, that instead of talking about individual numbers, we prefer instead to talk about whole sets of natural numbers. Do we have the resources to do *that*?

The answer is: It depends on the set in question. As we have seen, our system of logic allows for the use of predicates. We can define sets by looking for predicates that become true only when the variable is replaced by a number that is in the set. So long as the predicate involves only arithmetical notions, it will be expressible within Peano arithmetic.

For example, let us define $E(x)$ to be the predicate: "x is evenly divisible by 2." This predicate is purely arithmetical and is therefore intelligible within the system. Moreover, $E(x)$ is true precisely when x is replaced by an even number. In this way, we can say that the set of even numbers is definable within Peano arithmetic. In contrast, suppose that we define $S(x)$ to be the predicate: "x is written entirely with straight-line segments when expressed with the standard Arabic numerals." This predicate is not intelligible within the system. Peano arithmetic knows nothing of the particular symbols we use when writing numbers.

We now have sufficient background to comprehend the punch line, which is addressed in Chapter 10. Remarkably, it turns out that any formal system capable of expressing basic arithmetic (a system such as **PA**, for example) is also capable of expressing a truly remarkable sentence.

9.4 Notes and Further Reading

My purpose in this chapter has been to describe just enough of the technical minutiae to make the developments of the next chapter intelligible. The details of quantification and predication get technical in a hurry, which explains why even elementary texts in this area quickly devolve into an ocean of incomprehensible notation. Courageous readers wanting to know more about this theory can have a go at the books by Quine (1958), Smullyan (1995), and Smith (2012).

The whole subject of quantification theory, by which we mean the manner in which intuitive notions of quantity are captured in formal systems, has a long history. A helpful reference is the paper by Bonevac (2012).

For a readable discussion of axiomatic systems in general, try the small book by Blanché (1962).

CHAPTER 10

The Empuzzlement of Gödel's Theorems

Lewis Carroll empuzzled the central principles of Aristotelian logic, thereby making them accessible to an audience that would find the text-book treatments unintelligible. In so doing, he achieved something of great importance, independent of any debates regarding his contributions to scholarly treatments of logic.

Raymond Smullyan achieved a similar feat by empuzzling the two great incompleteness theorems of Kurt Gödel. The first theorem states, roughly, that in any formal system that is capable of expressing the basic arithmetic of natural numbers, there must be sentences that are true but unprovable. Gödel proved this in 1931, and it came as a shock to many mathematicians of the time. The second theorem asks whether it is possible, within such a formal system, to formulate and prove the statement "This system is consistent." The answer is that the statement can be formulated, but not proved, within the system.

There is much to unpack if we are to understand all of this. Be warned that this chapter is the most technically difficult chapter in the book, so some patience might be required. I believe you will find it is worth the effort.

10.1 Established Knights and Knaves

Since the solutions to the puzzles in this section are essential to the flow of our story, they have been printed right after the puzzles themselves. You might want to cover up these solutions at first, so that you can solve the puzzles yourself.

We now return to the island of knights and knaves. As before, assume that knights always tell the truth, while knaves always lie. On one especially interesting island, we find that some of the knights are said to be *established* knights, and some of the knaves are *established* knaves. Islanders bearing the label "established" can be thought of as having proved their identities.

Thus, there are four sorts of people living on the island: established and unestablished knights, and established and unestablished knaves. As a warm-up for what is to come, have a go at the following puzzle:

Puzzle 44. *Two islanders, whose names are Anne and Bart, approach you. Anne says, "I am not an established knight." Bart says, "I am an established knave." What can you conclude about Anne and Bart?*

Solution: If Anne were a knave, then her statement would be true. Since knaves do not make true statements, we can conclude that she is actually a knight. But then her claim that she is not an established knight is true. We conclude that she is a knight, but not an established one. By very similar reasoning, we conclude that Bart is an unestablished knave. A knight who made Bart's statement would be lying. So Bart is a knave, and since his statement is false he must be unestablished. □

Let us further suppose that the islanders have formed various social clubs. Any given islander might be a member of more than one club or might not be a member of any club. Interestingly, the clubs on the island satisfy certain conditions, which we shall label E_1, E_2, C, and G. Here are the conditions:

- E_1: The set of all established knights forms a club.
- E_2: The set of all established knaves forms a club.
- C: Given any club C, the set of all islanders who are not members of C form a club of their own. This club is referred to as the *complement* of C, and is denoted by $\overline{C}$.
- G: Given any club C, there is at least one inhabitant of the island who claims that he is a member of C. This is referred to as the *Gödelian condition*, for reasons that shall become clear later.

It turns out that these conditions have some interesting consequences.

Puzzle 45. *Prove that if all four conditions hold, then there must be at least one unestablished knight and also at least one unestablished knave. (Smullyan 1978, 226)*

Solution: Here is one way to solve Puzzle 45. By E_1, we know the set of established knights forms a club. Condition C now tells us that $\overline{E_1}$, which contains all of the islanders who are not established knights, is also a club. We now invoke G to conclude that there is at least one islander who claims to be a member of $\overline{E_1}$. In other words, he claims not to be an established knight. But this is just precisely the situation considered in Puzzle 44. It is only an unestablished knight who can claim not to be an established knight.

To show there is an unestablished knave, we begin with E_2, which tells us that the set of all established knaves forms a club. Condition G now ensures that there is someone who claims to be a member of this club, implying that someone on the island claims to be an established knave. But as we saw in the solution to Puzzle 44, it is only an unestablished knave who can make such a claim. □

Having worked through that puzzle, here is another one to ponder.

Puzzle 46. *Show that neither the set of all knights nor the set of all knaves can be a club on the island. (Smullyan 1978, 226)*

Solution: If the set of all knaves formed a club, then by condition G, there would be someone on the island who claims to be a member of this club. That is, this person would claim to be a knave, which is impossible. If the set of all knights formed a club, then condition C would guarantee that the set of all knaves would also be a club. But we have just seen that this is impossible.

As an aside, we note that the result of Puzzle 46 makes possible an alternative solution to Puzzle 45. To show that there is at least one unestablished knight, suppose instead that there were actually no unestablished knights. In this case, the set of all knights would be identical to the set of all established knights. But condition E_1 tells us that the set of all established knights forms a club. This is a contradiction, since we have just seen that the set of all knights does not form a club. Likewise, if there were no unestablished knaves, then the set of all knaves is the same as the set of all unestablished knaves. But E_2 says the set of all established knaves forms a club, while we have just seen that the set of all knaves is not a club. We have reached another contradiction. □

Taken by themselves, puzzles about established knights and knaves might seem like just another variation on a familiar theme. Smullyan, however, had higher ambitions for these teasers. Embedded in these puzzles are the central ideas underlying Gödel's first theorem.

10.2   A Sentence That Is True but Unprovable

In a formal system, some statements are true, and some statements are provable. Both of these concepts require some thought.

It seems clear enough what we mean by "provable." We are given certain axioms, and we are given certain rules for correct inference. If you can arrive at some statement by a sequential application of the rules to the axioms, then your statement is provable.

For example, in **PA**, the axioms tell us that 0 is a number and that $s(0)$ is also a number. Then the second addition axiom, applied to the case where $x = s(0)$ and $y = 0$, tells us that

$$s(0) + s(0) = s(s(0) + 0).$$

The first addition axiom now tells us that

$$s(s(0) + 0) = s(s(0)).$$

If we use the more standard notation that $s(0) = 1$ and $s(s(0)) = s(1) = 2$ (that is, the number after 0 is 1, and the number after 1 is 2), then we have just proved that $1 + 1 = 2$. In a sufficiently rich axiom system, chains of inferences can be strung together in astonishing ways to establish conclusions undreamed of when the system was devised. This fact is what keeps mathematicians in business.

Note, however, that in mathematics, there is no absolute notion for what it means for something to be provable. Instead, the notion of "provability" only makes sense in a specific formal system. Smullyan illustrates this idea dramatically by asking us to consider the sentence "This statement can never be proved." Let us refer to this sentence as P (Smullyan 2015, 50).

At first blush, sentence P seems to lead to a paradox, because any attempt to assign it a truth value leads to contradiction. If we decide that P is false, then it must be that P really *can* be proved. But if it can be proved, then P must be true, which is a contradiction. So, we have just proved that P cannot be false. Does that not entail that it is true? But in that case I have, indeed, proved that P is true, which is contrary to what it plainly says. I get a contradiction either way.

Smullyan now notes the problem is that the notion of "provable" is not well defined in this context. This concept is only meaningful in the context of a properly specified formal system. So, let us imagine that we have described some formal system, call it S, within which we can discuss sentence P. In that case, we would have to revise P so that it says, "This statement can never be proved in system S." But there is nothing at all paradoxical in *that* statement.

So "provable" seems clear enough, at least in the context of a given formal system. However, "truth" within a formal system is a far murkier notion. Throughout this entire discussion, I have been at pains to emphasize that a formal system is just that: formal. The symbols have no meaning outside the system itself, and the axioms are just the rules of the game. But our everyday notion of "truth" is all about correspondence with reality. What, then, could it possibly mean to say that a statement is true in a formal system?

The answer lies in the concept of a *model* for the system. Though I have emphasized the formality of our formal systems, I have also emphasized that they do not arise out of thin air. Instead, they are generally designed with a specific interpretation in mind. In our discussion of the axioms for Euclidean geometry, we knew perfectly well what points and lines were long before we looked at any formal systems. Likewise for **PA**. No one first learns about the natural numbers by studying Peano's axioms. Instead, people come to recognize the reasonableness of the axioms by first understanding the natural numbers.

This is how we come to understand the notion of truth in a formal system. We can construct a model for the system, by which we mean an interpretation of the undefined terms and their relationships, in which the axioms are all seen to be true. For example, the set of natural numbers can be viewed as a model for Peano's axioms. In a model, it makes perfect sense to say that a statement is true or false. In the natural numbers, it is true that 2 is the only even prime, and it is false that 16 is an odd number. No mystery here. Now, if a statement must be true in any model of the formal system, then we can reasonably say the statement is true in the system.

Thus, in a formal system, there are true statements and there are provable statements, but those two notions are distinct. Of course, in practice, truth and provability go together. We would not accept that a statement about the natural numbers is true unless it were backed up by a proof. But that is irrelevant for our present purposes. We generally take it for granted that mathematical statements have truth values, regardless of whether we know what those truth values are.

And now, finally, we are ready to discuss what Gödel accomplished. On one hand, we have the set of all true statements. On the other, we have the set of all provable statements. They are conceptually distinct, but we might suspect that in practice, they are the same. We might hope that within a formal system, any true statement is provable, and any provable statement is true. Remember that the whole point of working in a formal system is put some area of mathematics on a solid, logical footing. If there are true statements that are not provable, then our system fails to capture all of the richness of the mathematics in question. It is even worse if the system can actually prove false statements. Such a system would be mathematically useless.

A system in which every true statement is provable is said to be *complete*. A system in which it is impossible to prove a false statement is said to be *consistent*. Ideally, we want a system that is both complete and consistent.

But Gödel's first theorem shows that such a thing is impossible. If your system is consistent, and if it is powerful enough to express basic arithmetic, then it must be incomplete.

In Section 9.2, we considered the question of what can and cannot be asserted in a particular formal language. Such considerations lie at the heart of Gödel's proof. To see the main idea, suppose that your formal system is capable of proving, of some sentence Q, that "Q is true if and only if Q is unprovable." What would be the consequences of this?

Well, suppose that the system is consistent. That means it cannot prove anything that is false. Could it then be that Q is provable? If it is, then the second part of the biconditional is false. But since the system is consistent, and since it proves the biconditional, we conclude that the biconditional as a whole is true. This implies the first part must be false as well (since a biconditional is true only when both of its parts have the same truth value). In this case, we would conclude that Q is false. In other words, Q is false, but provable. This implies the system is inconsistent after all, which is a contradiction.

We must therefore conclude that Q is unprovable. But in this case, for the biconditional as a whole to be true, we must have that Q is true. Therefore, Q is true, but unprovable. This immediately implies that the system is incomplete.

A consistent system that is sufficient for expressing basic arithmetic must be incomplete. Put differently, such a system will either prove something that is false, or else it will contain true statements that cannot be proven.

Take your pick.

10.3 Establishment, Revisited

Gödel showed that any system capable of expressing arithmetic is also capable of expressing a sentence playing the role of Q.

When you first hear this, it sounds utterly incredible. The language of Peano arithmetic is well suited to making statements about natural numbers. But how on earth can it make assertions about what is provable and what is not?

The trick is to establish a correspondence between the well-formed formulas of the system (its sentences) and natural numbers. It turns out that it is possible to assign a number to each formula in such a way that two things are true:

(1) Every formula gets its own number.

(2) The formula can be recovered simply by examining its number.

To make a simplistic analogy, suppose I number the letters of the alphabet from 1 to 26, with A being number 1 and Z being number 26. (We will not make a distinction between capital and lowercase letters). Let the number 0 represent a break between letters, and let 00 represent a break between words. With this scheme in place, I can represent the phrase "logic is fun" by the number

$$12015070903009019006021014.$$

Notice that L is the twelfth letter of the alphabet, O is the fifteenth, and so on. Given only that 26-digit number, I could recover the phrase (though the recovery process would be rather tedious.) This sort of thing can be done with the formulas in a formal system.

A numbering of this sort, in which formulas of a system are assigned unique natural numbers, is referred to as a *Gödel numbering* for the system. In this manner, assertions about formulas can be construed as statements about natural numbers. Moreover, this idea is readily extended to sequences of formulas, by concatenating the individual Gödel numbers of the formulas comprising the sequence (with a special symbol reserved for the break between formulas). This idea then extends to proofs, in the sense that every proof can be said to receive its own Gödel number. After all, a proof is just a particular sort of sequence of statements (one in which every statement is either an axiom or a formal consequence of applying the deduction rules to the axioms).

It is in this manner that it becomes possible to make our formal language, which is designed for making statements about natural numbers, make statements about provability.

Let me be more specific. Once the formulas of the system have received a Gödel numbering, the following predicate becomes intelligible: "The number x is the Gödel number for a proof of the sentence whose Gödel number is y." If you replace x and y with specific numbers, then it is possible in principle to determine whether the resulting sentence is true. We simply retrieve the sequence of formulas represented by x and check to see whether the sequence comprises a valid proof of y, in much the same way that a math teacher would check to see whether a student's work is correct. If y is provable, then there is *some* number x that is the Gödel number for a proof of y.

Thus, we can define a predicate $E(y)$ as follows: "y is the Gödel number of a sentence for which there exists a number x that is the Gödel number for a proof of y." This predicate is true precisely when y is replaced by the Gödel number of a provable sentence.

We have just shown that the set of provable sentences is definable in our system.

This brings us back to our established knights and knaves. Smullyan intends for us to construe the set of all knights to be the set of all true statements and the set of all knaves to be the set of all false statements. The sets of established knights and knaves should be thought of, respectively, as the set of statements that are provably true and provably false.

Furthermore, armed with a Gödel numbering for its formulas, the system of Peano arithmetic has the resources to formulate the equivalents of Smullyan's conditions E_1, E_2, C, and G (see Section 10.1). More specifically, the set of provably true statements can be defined in the system, as can the set of provably false statements. This covers conditions E_1 and E_2. Condition C is equivalent to the idea that if a set is definable in the system, then so is the complement of that set.

Condition G is more difficult. Recall that the puzzle version was, "Given any club C, there is at least one inhabitant of the island who claims that he is a member of C." The translation from puzzles to Peano arithmetic proceeds by analogizing clubs to sets of numbers and inhabitants of the island to formulas in the system. Condition G can be taken to state that for any definable set in the system, there is a formula that asserts, for some number n, that "n is the Gödel number for a formula in the set."

The proof that there is an unestablished knight on the island is analogous to proving that there is a statement that is true but unprovable in the formal system. In the puzzle version, the proof proceeded quickly:

- The set of established knights forms a club.
- Therefore, the set of people who are not established knights forms a club.
- So there is an islander who claims to be a member of the club of unestablished knights.
- But only an unestablished knight can make such a claim.

The key to this argument is the self-reference in bulleted item three. Smullyan's four conditions (though technically we do not need E_2 for this part of the argument) guarantee that there is a knight who will speak about himself in interesting ways.

In the formal system version, we can recast the argument like this:

- The set of provable statements is definable in the system.
- Therefore, the set of statements that are not provable is definable.
- So there is a sentence S that says, "n is the Gödel number of a sentence that is not provable in the system."
- But this can be done in such a way that n is actually the Gödel number of S itself.

We are then left with two options. If S is true then, as it says, it must not be provable in the system. If it is false, then what it says is not the case, implying that it is provable after all. In the first case, we have a sentence that is true but unprovable. In the second case, we have a sentence that is false but provable. If we assume the system is consistent, then the second case is ruled out.

In other words, if the system is consistent, then there must be a sentence that is true but unprovable.

10.4 A Gödelian Machine

Of course, if Gödel's theorem were really as easy to prove as Smullyan's puzzles make it appear, then it would not today be celebrated as a milestone in the history of logic. Certainly, the early stages of Gödel's proof are straightforward enough. The mechanics of assigning the Gödel numbers is not really so terrible, and neither is the idea of turning "provability" into an arithmetic predicate. The third step, where we show it is possible to express a statement exhibiting the proper sort of self-reference, is far more complex. Smullyan suppressed all relevant details by building them into condition G.

He did, however, devise another puzzle that helps illustrate the sort of self-reference employed in the proof of Gödel's theorem (Smullyan 1992b, 2–3). He asks us to imagine a machine that prints out certain expressions. These expressions are made up entirely of five symbols:

$$P \quad N \quad \sim \quad (\quad).$$

By an *expression* we mean any finite sequence using only these five symbols. Here are some examples of expressions:

$$P(\sim NN, \qquad ()(\sim P, \qquad PPN(() \sim))P.$$

For now, these expressions are just nonsense strings, but we will soon single out certain expressions to be given a meaning. Let us also imagine that the machine is programmed in such a way that any expression it is capable of printing is eventually printed.

We will say that an expression is *printable* if the machine is capable of printing it. Also, given any expression X, we define the *norm* of X to be the expression $X(X)$. For example, the norm of the expression "$P\sim$" is the expression

$$P \sim (P \sim),$$

and the norm of "*PPNN*" is

$$PPNN(PPNN).$$

At this point, we can assign a meaning to certain expressions. Let us interpret *P* to stand for "printable," *N* to stand for the norm of an expression, and ~ to indicate "not." With this interpretation, certain expressions become intelligible. For example, we would take $P(X)$ to be saying "*X* is printable," where *X* represents an expression. The expression $PN(X)$ says, "The norm of *X* is printable," and ~$PN(X)$ means, "The norm of *X* is not printable."

Let us, then, define a *sentence* to be an expression of one of the following four types:

$$P(X), \qquad PN(X), \qquad {\sim}P(X), \qquad {\sim}PN(X).$$

Given any expression X, the sentence $P(X)$ is true if *X* really is printable, and the sentence $PN(X)$ is true if the norm of *X* is printable. Likewise, ~$P(X)$ and ~$PN(X)$ are true, respectively, when *X* is not printable or when the norm of *X* is not printable.

We are now given one more piece of information: It turns out that the machine is always accurate, in the sense that it only prints true sentences. If it ever prints out $P(X)$, then you can be certain that *X* really is printable, and if it ever prints out $PN(X)$, then you can be certain that $X(X)$ is printable.

The machine exhibits an interesting sort of self-reference, in that it prints sentences that make assertions about what the machine is able to print. Now, we are given that the machine will never print something that is false, but we might wonder whether it is capable of printing out all true sentences. It turns out that the answer is no, and that leads us to our next puzzle.

Puzzle 47. *Find a sentence that the machine cannot print.*

Having considered the proof of Gödel's theorem, we might suspect that the answer lies in finding a sentence that asserts its own unprintability. But how are we to do that?

Solution: Here is a sentence that is true, but unprintable: ~$PN({\sim}PN)$. Let us refer to this sentence as *S*.

The first three symbols of *S* say, in effect: "The norm of the thing in the parentheses is not printable." The thing in the parentheses is the expression ~*PN*. But then the norm of the thing in the parentheses is ~$PN({\sim}PN)$, which is precisely *S* itself.

Now suppose that S were printable. Since the machine eventually prints anything it is capable of printing, it will eventually print S. And since the machine only prints true sentences, we would conclude that S is true. But S says that it is not printable, which is plainly false, given that the machine has, in fact, printed it. This is a contradiction.

We have shown that the machine cannot print S. More precisely, we have shown that "The norm of $\sim PN$ is not printable." This is exactly what S said. So, we have shown that S is true, but unprintable. □

10.5 Gödel's Second Incompleteness Theorem

In Section 10.2, we identified two important properties that a formal system may or may not have: completeness and consistency. We said that a formal system is complete if any true statement in the system is provable, and that it is consistent if it is impossible to prove a false statement. Gödel's first theorem addressed completeness. Informally, it told us that any formal system useful for mathematics will necessarily be incomplete. The key to doing so was the realization that any formal system capable of expressing arithmetic also has the resources to make assertions about provability.

We might wonder whether any such formal system can also make assertions about its own consistency. Can we express the statement "This system is consistent," and if so, is it possible to prove that the statement is true while working entirely in the system itself? Less formally, can a system prove its own consistency? Gödel resolved these questions as well. His answers were: Yes, we can formulate the statement, but no, we cannot prove it in the system.

The first part is straightforward. Let S denote any sentence that can be proven false in the system. Gödel used the statement $1 = 0$ for this purpose. If S is provable, then the system is inconsistent, and if the system is inconsistent, then S is provable. (This is because anything is provable in an inconsistent system, as we shall discuss in Section 12.1.) Thus the system is consistent if and only if S is not provable. Now, we have seen that the statement "S is not provable" can be construed as arithmetic, and is therefore expressible in the system. Moreover, we can take "S is not provable" to be asserting that the system is consistent, in the sense that the two statements are logically equivalent.

This shows that the system can be made to assert its own consistency. However, this assertion cannot be proved, for the following reasons.

Let T represent a sentence that asserts its own unprovability, as discussed in Section 10.3. Then the statement $S \rightarrow T$ says: "If the system is consistent, then T is not provable." Gödel's first theorem tells us that this statement is true. Let us suppose for a moment that this statement is not merely true but also provable. What would follow from this assumption?

The situation would then be that $S \to T$ is provable, but T is not provable. What about S? If S is also provable, then the statements S and $S \to T$ would constitute a proof of T. But proving T entails that the system is inconsistent. So, if the system is consistent, then S is not provable. A system cannot prove its own consistency.

This was all contingent on the assumption that $S \to T$ can actually be proven in the system. It turns out that this assumption is correct, though proving it is well beyond the scope of this book.

These ideas all take some getting used to, especially if you are unaccustomed to working with formal systems. Happily, Smullyan devised an ingenious collection of puzzles that illustrate the main ideas involved (Smullyan 1987, 67–106). These puzzles involve knights and knaves, of course. Smullyan's idea was to speak of reasoners and what they believe, as opposed to formal systems and what they can prove.

Let us start with a warm-up puzzle.

Puzzle 48. *A visitor to the island of knights and knaves meets an islander who says, "You will never know that I am a knight." What can we conclude from this?*

Solution: The visitor might reason as follows: "Suppose the native is a knave. Then his statement is false. That means that at some point, I *will* know that he is a knight. But this is impossible, since I cannot know something that is false. So the hypothesis that the native is a knave leads to a contradiction."

The visitor will now continue like this: "I have just proven that the native is a knight. That is, I now *know* that he is a knight. In this case, the native's statement is false. But this is another contradiction, since knights cannot make false statements." At this point, the visitor concludes that we have described a paradoxical situation. □

What do you think of that argument? Should we conclude that the puzzle has described an impossible situation? Smullyan notes that our solution to the puzzle is not really wrong, but it *does* hinge on a number of assumptions that have not been made explicit.

Note first that our analysis of the puzzle did not hinge solely on formal logic; it also depended on what the visitor knew. We could imagine frivolous cases—the native was speaking to a corpse, the native was speaking to a deaf person—which plainly do not give rise to anything paradoxical. So we certainly must assume that the visitor was alive and heard what the native said.

But soon we come to less frivolous concerns. If the visitor is to reason in the manner suggested by our solution, then it must be that he has a certain amount of reasoning ability. That is, we must assume he has some facility for logical argument. We also have to assume that the visitor not only understands the rules of the island—that knights only make true statements, knaves only make false statements, and everyone is either a knight or a knave—but also believes those rules. After all, if a visitor really confronted this situation and understood that a paradox ensued from his analysis of what the native said, he would probably just assume he had been misinformed about the nature of the island.

Let us stipulate that the visitor reasons flawlessly and believes the rules of the island. There is still a subtle point to consider. In principle, it is possible to know something without realizing that you know it. Our visitor went straight from "The assumption that the native is a knave leads to a contradiction" to "I now know that the speaker is a knight." This step is legitimate only if we assume that when the visitor knows something, he actually knows that he knows it.

If we accept all these assumptions, then we do, indeed, have a paradox. If the visitor's mental attributes are as we have described them, then it is not possible for the native to say to him, "You will never know that I am a knight." (Of course, we also have to assume that the native *knows* the mental attributes of the visitor, but we will not worry about that. We simply assume that the natives know everything they must know to behave in accordance with the rules of the island.)

Now let us move on to the main event. Instead of discussing what a visitor might *know*, let us consider instead some things he might *believe*. We want to avoid very exotic situations where, say, a visitor believes something but also believes that he does not believe it. In particular, let us suppose the visitor has the following mental attributes (note that a *tautology* is a statement like $P \vee \neg P$ that must be true as a matter of logic):

(1) He is well versed in logical reasoning. This means he believes all tautologies, and if he ever comes to believe both P and $P \rightarrow Q$, then he will also believe Q.

(2) Moreover, he believes that believing both P and $P \rightarrow Q$ entails a belief in Q.

(3) If he ever believes a proposition P, then he also believes that he believes it. In fact, he believes that believing P entails that he believes that he believes it.

Let us make things more concrete. Suppose such a person visits the island and that a native makes a statement to him. Then the visitor will believe the statement if and only if he also believes the native is a knight. Furthermore, if he ever believes the native is a knight, then he will believe the statement as well. In addition to this, the visitor will also know that if he ever believes both P and ¬P, then he will be inconsistent.

We can now consider the main puzzle:

Puzzle 49. *Suppose the visitor believes himself to be consistent. He never simultaneously believes a statement and its negation. If a native says to him, "You will never believe that I am a knight," what can we conclude?*

Solution: We can conclude that the visitor is actually inconsistent!

The visitor will reason as follows:

- Suppose that I ever believe that the native is a knight. Then since I believe this will imply that his statement is true, I will believe that I do not believe that he is a knight.
- Whenever I believe something, I also believe that I believe it (as stipulated in item 3). Thus, since I believe the native is a knight, I also believe that I believe it.
- This shows that if I ever believe the speaker is a knight, then I will simultaneously believe and not believe that he is a knight. This would imply that I am inconsistent.
- But since I also believe that I am consistent, I must conclude that I will never believe the native is a knight. But this is precisely what he said! He said that I would never believe that he was a knight, and what he said was true, so he really is a knight.

At this point, the visitor believes that the native is a knight, and also believes that he believes it. He will now continue his reasoning like this: "I now believe he is a knight. He said I would never believe this. Therefore, he made a false statement and must not be a knight."

Thus, the visitor believes both that the native is a knight and also that the native is not a knight. So he is inconsistent. □

If the visitor believes that he is consistent, then he must actually be inconsistent. This implies that if the speaker really is consistent, then he cannot know that he is.

The analog for formal systems is that if a system is consistent, then it cannot prove that it is consistent. Put differently, if a system proves its own consistency, then it must actually be inconsistent.

This is the substance of Gödel's second theorem.

10.6 Puzzles for Solving

We have trod some difficult ground to reach this point, so let us close with some puzzles for you to mull over. First, there are a few about established knights and knaves. Recall that on the knight/knave islands in question, some knights are *established* knights and some knaves are *established* knaves. Moreover, in Section 10.4, I listed four characteristics that such an island might possess:

- E_1: The set of all established knights forms a club.
- E_2: The set of all established knaves forms a club.
- C: Given any club C, the set of all islanders who are not members of C form a club of their own. This club is referred to as the *complement* of C, and is denoted by $\overline{C}$.
- G: Given any club C, there is at least one inhabitant of the island who claims that he is a member of C. This is referred to as the *Gödelian condition*.

Puzzle 50. *Suppose that a particular island satisfies conditions E_1, E_2, and C (but not necessarily G). Suppose further that it possesses a new property, which we shall call GG:*

- *GG: Given any two clubs C_1, C_2, there are islanders A and B such that A claims that B is a member of C_1, and B claims that A is a member of C_2.*

Do the knights of the island form a club? Do the knaves? Can you determine whether there are any unestablished knights or unestablished knaves on the island? (Smullyan 1978, 230)

Puzzle 51. *Now suppose you visit an island on which conditions E_1, E_2, and GG all hold, but not necessarily conditions C or G. This time you want to prove that there is either an unestablished knight on the island or an unestablished knave. Also prove that it is impossible that both the knights form a club and the knaves form a club. (Smullyan 1978, 232)*

Let us move now to puzzles involving Gödelian machines of the sort we saw in Section 10.4. Smullyan has explored puzzles of this sort in several of his books, most notably in *The Lady or the Tiger, and Other Logic Puzzles*, and *The Gödelian Puzzle Book*. The point of these puzzles is that his machines have properties that mimic certain aspects of the logical systems studied by Gödel. Thus, the puzzles eventually build up to showing how Gödel was able to craft a sentence that referred to its own provability, starting from a system that did not seem capable of expressing such a thought. However, these puzzles quickly become so intricate and difficult that I will only present a few of the simpler ones here.

These puzzles require some setup, so let me turn the floor over to Smullyan. In what follows, a character named Arnold is describing to the first-person narrator a strange machine he has built.

"This device operates on expressions consisting of strings of capital letters," [Arnold] said. "Any string of capital letters is called an *expression*. An expression doesn't have to be meaningful. For example, BLPQ is an expression and so is AHM—also any capital letter itself is an expression. Now, one feeds an expression into the gadget, and if the expression is acceptable, the machine then outputs an expression—usually different from the input. I shall use small letters x, y, z as standing for arbitrary expressions. . . . And for any pair of expressions x and y, by xy I mean the expression x followed by the expression y—for example, if x is the expression BLD and and y is the expression HIJZ, then xy is the expression BLDHIJZ. Is that clear?"

"Yes," I said, "Go on!"

"I shall say than an expression x *yields* an expression y—in short 'x yields y'—to mean that if x is fed in as input, y comes out as output."

Arnie then told me two rules concerning the operation of the gadget. He first mentioned his Rule Q:

Rule Q: For any expression x, the expression Qx yields x.

"For example, QBLZ yields BLZ," said Arnie. . . .

"Before I state the second rule," said Arnie, "I must tell you that for any expression x, the expression xQx . . . is known as the *companion* of x. My second rule is this:"

Rule C: If x yields y, then Cx yields the companion yQy of y.

"For example, since QAB yields AB, then CQAB should yield the companion of AB, which is the expression ABQAB—here, let me try it."

Arnie then fed in CQAB, and sure enough, out came ABQAB.

"My gadget never fails!" said Arnie, with a proud smile. "Indeed, for any expression x, CQx will yield the companion of x." (Smullyan 2013, 86)

Now for some puzzles:

Puzzle 52. *Find an x that yields itself (you can feed in x, and x comes out). Then find an x that yields its own companion. (Smullyan 2013, 87)*

Puzzle 53. *Prove that for any expression y, there is some x that yields yx. For example, what x yields Ax? (Smullyan 2013, 87)*

Puzzle 54. *Now let us make a new definition: For any expression x, the repeat of x is the expression xx, that is, x followed by itself. Let us also introduce a new rule:*

Rule R: *If x yields y, then Rx yields the repeat of y.*

Given this rule, find an x that yields its own repeat. Then find an x that yields the repeat of its companion. (Smullyan 2013, 87)

I close with some examples in the vein of Smullyan's empuzzlement of Gödel's second theorem. Smullyan considered a variety of different types of reasoners, differentiated by the things they believe. His point was to model certain questions about formal systems and the statements provable in them in terms of reasoners and the sorts of things they believe.

Toward that end, let us suppose that all of our reasoners are flawless logicians. More precisely, they believe all tautologies, and if they ever believe a proposition P and also believe that $P \rightarrow Q$, then they will believe Q.

Now, for any proposition P, denote by BP the proposition that the reasoner believes P. Let us say that a reasoner is *normal* if for any proposition P, if he believes P, then he believes that he believes P. (In other words, if he believes P, then he also believes BP.) Finally, let us say that a reasoner is *regular* if any time he believes a proposition of the form $P \rightarrow Q$, he also believes $BP \rightarrow BQ$.

With this setup, try your hand at a few puzzles.

Puzzle 55. *Show that if a regular reasoner believes that P is logically equivalent to Q, then he will also believe that BP is logically equivalent to BQ. (Smullyan 1987, 92)*

Puzzle 56. *Suppose that Joe is a regular reasoner, and that there is at least one proposition Q for which Joe believes BQ. Show that this implies that Joe is normal. (Smullyan 1987, 92)*

10.7 Notes and Further Reading

Gödel's theorems and the ideas underlying their proofs certainly take some getting used to! The biography of Kurt Gödel written by Rebecca Goldstein (2006) provides helpful historical context and also takes a stab at explaining the theorems themselves. For readers with some mathematical background, the book by Nagel and Newman (2001) is worth a look. Readers with a lot of mathematical background can take a shot at Smullyan's own textbook (Smullyan 1992b).

There is no question that Gödel's theorems were ingenious in their approach, and that they forced a major rethinking of certain ideas that were then prevalent in the philosophy of mathematics. However, it is also easy to exaggerate their importance. Most mathematicians woke up the day after the publication of Gödel's work and approached their craft as though nothing had happened. However, the theorems spawned a veritable industry of writers attempting to apply them in domains far removed from obscure questions about formal systems. This includes attempts to use them in the service of proving that God exists. Much of this work, it must be said, was written by crackpots. For a survey of the uses and abuses of the theorems, I recommend the book by Franzén (2005). The paper by Edis (1998) is also worth a look.

Smullyan's scholarly work involved questions surrounding self-reference and Gödel's theorems. The recent volume edited by Fitting and Rayman (2017) contains many interesting essays about this work. The anthology edited by Rosenhouse (2014) contains additional essays about and tributes to Smullyan's work.

10.8 Solutions

Puzzle 50. We can show that neither the set of all knights nor the set of all knaves forms a club. According to condition C, if either the set of knights forms a club or the set of knaves forms a club, then the other would as well. In other words, either both are clubs or neither is a club. Suppose they are both clubs. Then by condition GG, there are islanders A and B such that A says, "B is a knave," and B says "A is a knight." But this situation is paradoxical. Therefore, we conclude that neither the set of knights nor the set of knaves forms a club.

Since the set of knights is not a club, but the set of established knights is a club, it follows that the two sets are different. Likewise, the set of knaves is different from the set of established knaves. It follows that there is at least one unestablished knight and at least one unestablished knave.

This solution gets the job done, but Smullyan notes that there is a more complex, but also more illuminating, approach to the puzzle. By property E_1, the set of established knights forms a club. By property C, the set of people who are not established knights also forms a club. Denote these clubs by C_1 and C_2. By property GG, there must be islanders A and B who assert:

$A : B$ is an established knight.

$B : A$ is not an established knight.

Now, suppose that A is a knight. Then his statement is true, which implies that B really is an established knight. Therefore *his* statement is true as well. Therefore, A is an unestablished knight. In contrast, if A is a knave, then he is certainly not an established knight. This implies that B's statement is true, and therefore that B is a knight. But since A's statement is false in this scenario, it must be that B is an unestablished knight.

Either way, there is an unestablished knight.

The proof that there is an unestablished knave is very similar. The established knaves form a club, as do the islanders who are not established knaves. Therefore, there are islanders who say

> $A : B$ is an established knave.
>
> $B : A$ is not an established knave.

An analysis nearly identical to that above now shows that there must be an unestablished knave.

Puzzle 51. The solution to this puzzle is similar to that of Puzzle 50, but it also has some interesting differences. If the knights and knaves were both clubs, then there would be islanders A and B such that A says that B is a knave, and B says that A is a knight. This is paradoxical, as we have seen before. Thus it cannot be that the set of knights and the set of knaves are both clubs. Either one by itself can be a club, but not both. Proceeding as before, we see that either the set of knights differs from the set of established knights (in that one is a club and the other is not), or the set of knaves differs from the set of established knaves. Thus, there is either an unestablished knight or an unestablished knave, though we do not know which.

Also as in Puzzle 50, there is a more interesting approach to the problem. Since the established knights and established knaves both form clubs, there are islanders A and B who say:

> $A : B$ is an established knave.
>
> $B : A$ is an established knight.

Now, suppose A is a knight. Then B really is an unestablished knave. That makes B's statement false, which is possible only if A is a knight, but unestablished. Likewise, if A is a knave then B's statement is again false. Thus B is a knave. But since A's statement is false in this scenario, it must be that B is a knave, but unestablished.

This reasoning shows that there is either an unestablished knight or an unestablished knave, but once more we cannot tell which.

Puzzle 52. We know that for any x, we have that Qx yields x. It follows that CQx yields xQx. But what happens if x is C? Then we would find that CQC yields CQC, and this is the solution to the first problem.

To find an expression that yields its own companion, reason as follows: Since CQx yields the companion of x, we know that CCQx yields the companion of the companion of x. Now take CC for x, and so CCQCC yields the companion of CCQCC. Thus, CCQCC yields its own companion.

Puzzle 53. We know that QAC yields AC. Therefore, CQAC yields the companion of AC, which is ACQAC. Thus, the expression CQAC yields an A followed by the very expression CQAC.

We can generalize this principle to say that for any expression y, an x that yields yx is CQyC. This is because CQyC yields the companion of yC, which is yCQyC. So, CQyC yields the expression y followed by the expression CQYC.

Puzzle 54. As before, we know that for any x, the expression CQx yields the companion of x. If we let x be RC, then we get that CQRC yields the companion of RC, which is RCQRC. It follows that if we put an R in front of CQRC, then we will get the repeat of RCQRC. So, RCQRC yields its own repeat.

Now we seek an x that yields the repeat of its companion. As always, CQx yields the companion of x. Taking x = RCC, we get that CQRCC yields the companion of RCC, which is RCCQRCC. Thus, if we now put an R in front of CQRCC, we will get the repeat of RCCQRCC, which is the repeat of the companion of RCC.

Puzzle 55. Suppose the reasoner believes that P is logically equivalent to Q. Then he must believe both P → Q and Q → P. Since he is regular, he will also believe BP → BQ and BQ → BP. But this entails that he will believe that BP is logically equivalent to BQ.

Puzzle 56. Recall that a regular reasoner is one who, if she believes that P → Q, will also believe that BP → BQ. A normal reasoner is one who, if she believes P, will also believe BP.

Now, suppose Joe is a regular reasoner. Also suppose that Q is some proposition for which Joe believes BQ. Now let P be some other proposition that Joe believes. We need to show that Joe also believes BP.

The proposition P → (Q → P) is a tautology. Hence, Joe believes it. Moreover, we are assuming that he believes P. It follows that he will believe Q → P. Since he is regular, it now follows that he will believe BQ → BP. But since we are assuming that he believes BQ, it follows that he will believe BP, which is precisely what we wanted to show.

CHAPTER 11

Question Puzzles

In his doctoral thesis of 1867, mathematician Georg Cantor wrote: "In mathematics, the art of asking questions is more valuable than solving problems." In this chapter, we show that in logic, at least sometimes, this maxim remains true. We will consider knight/knave puzzles in which the goal is the formulation of questions so clever that the answers are informative, even if you do not know the types of the people to whom you are speaking. Along the way, we shall develop some general principles for handling such puzzles.

Throughout this chapter, we will continue to refer to knights and knaves. Recall that knights always tell the truth, and knaves always lie. Since the solutions to the puzzles are integral to the points I wish to make, they are presented immediately after the puzzles themselves.

11.1   Three Warm-Ups

Here is a straightforward representative of the sort of puzzle I have in mind:

Puzzle 57. *You meet triplets named Leon, Larry, and Tim. They are visually indistinguishable, but Leon and Larry are knaves, while Tim is a knight. (Note the first initial L for the liars and T for the truthtellers.) What one question could you direct to one of the brothers to determine whether or not he is Larry? (Smullyan 2009, 6)*

Solution: We might try asking simple questions first. Suppose we ask one of the brothers, "Are you Tim?" This will get us nowhere, since everyone will reply yes. Tim will do so truthfully, while Leon and Larry will do so falsely. The question, "Are you Larry?" does a little better. Tim and Larry will both say no, but Leon will say yes. A "yes" answer definitively identifies Leon, but a "no" answer tells us only we are talking to Tim or Larry.

The interesting point of this puzzle, however, is that "Are you Leon?" gets the job done. Tim will tell the truth and say no, while Leon will lie and say no. But Larry will lie and say yes. So, a "no" answer implies you are not addressing Larry, while a "yes" answer implies that you *are* addressing Larry. Very nice! ☐

Here are two further warm-ups, conveniently grouped into one:

Puzzle 58. *Arthur and Robert are twin brothers, indistinguishable in appearance. One of the two always lies, and the other always tells the truth, but you are not told whether it is Arthur or Robert who is the liar. You meet one of the two one day and wish to find out whether he is Arthur or Robert. You may ask him only one yes/no question, and the question may not contain more than three words. What question will work? Alternatively, suppose instead that you wish to know whether it is Arthur or Robert who is the truthful one. What three-word question will work this time? (Smullyan 2009, 6)*

Solution: Smullyan notes an amusing symmetry between the two problems. To determine which is Arthur, you ask: "Is Arthur truthful?" To determine whether Arthur is truthful you ask: "Are you Arthur?" Let us consider the possibilities.

Suppose you ask: "Is Arthur truthful?" Arthur will definitely reply "yes," truthfully if he really is truthful, but falsely if he is not truthful. How will Robert reply? If he is truthful, then Arthur is not truthful. So Robert will truthfully reply "no." If Robert is not truthful, then Arthur is, meaning that the correct answer is yes. But since Robert is not truthful, he will lie and say "no."

Now suppose we ask, "Are you Arthur?" If he really *is* Arthur, then he will answer "yes" if he is truthful and "no" if he is not. If he is Robert, then he will answer "no" if he is truthful (implying that Arthur is not truthful), and he will answer "yes" if he is not truthful (implying that Arthur is truthful). Either way, a "yes" answer implies that Arthur is truthful, while a "no" answer implies that Robert is truthful. ☐

11.2 The Power of Indexical Questions

In Section 7.4, we discussed a variation on the basic knight/knave scenario in which inhabitants of an island can be either sane or insane. If a person is sane, then everything he believes to be true really is true. That is, they are correct in all of their beliefs. With an insane person the situation is reversed: They are incorrect in all their beliefs.

Let us review the implications of this. Suppose you ask someone if cats are mammals.

- A sane knight will know that the correct answer is yes, and that is what he will say.
- A sane knave will also know that the correct answer is yes, but he will lie and say "no."
- An insane knight will believe that the correct answer is no, and that is what he will say.
- An insane knave will believe that the correct answer is no, but he will lie and say "yes."

The point is that sane knights and insane knaves will always give the same answer to a given question. This might seem to make it impossible to distinguish them merely by asking them questions, which turns out to be incorrect, for a subtle reason. Puzzle 59 is adapted from Smullyan (1983, 14–15).

Puzzle 59. *Suppose you meet a pair of identical twins. It is known that one is a sane knight, while the other is an insane knave. Is there any number of yes/no questions you can ask one of them to determine which one he is?*

Solution: The trick is to ask a question that means different things to different people. Suppose we ask one of the brothers: "Are you a knight?" A sane knight will know that he is a knight, and will truthfully answer "yes." An insane knave will believe that he is a knight, but will then lie and reply "no." Just like that, we have distinguished between them.

Note that the question "Are you sane?" also works, but in this case, I leave the analysis to you.   □

Questions like the one used in this puzzle, which change their meaning depending on the context in which they are asked, are said to be *indexical*. It is true that a sane knight and an insane knave will give the same reply to any basic question of fact, of the sort that any normal person might ask. But a question can change its meaning when directed to different people. In this way, it is possible to direct the same collection of words to two different people, while nonetheless asking them different questions.

11.3 The Heaven/Hell Puzzle

In the field of question puzzles, there is one that is so famous, it towers over all others. What you are asked to do seems impossible, and the solution is so ingenious and satisfying, that the problem is an absolute, slam-dunk classic among brainteasers. Here is that puzzle:

Puzzle 60. *You are in a room with two unmarked doors. One leads to heaven, the other to hell. You must choose one of the doors. Of course, you would prefer the door that leads to heaven. Each door has its own guard, and the guard knows which door he is guarding. One of the guards is a knight, the other is a knave. You do not know which is which, however. You may ask a single yes/no question of one of the guards, after which you must make your choice. What should you ask?*

For convenience, let us refer to this as the heaven/hell puzzle.

We quickly realize that the normal sort of question that arises in daily conversation will get us nowhere. We could point to one of the doors and ask: "Does this door lead to heaven?" But what does *that* get us? If it really is the door that leads to heaven, then the knight will say yes and the knave will say no, and we will have no idea how to interpret the answer. If it is the door that leads to hell the answers will be reversed, but we will confront the same problem.

It will take a very clever question to evade this difficulty.

Solution: You point to one of the doors and ask either of the two, "If I asked the other person if this is the door to heaven, would he say 'yes'?"

Suppose you really are pointing to the door to heaven. In that case, when asked whether it is the door to heaven, the knight would say "yes," while the knave would say "no." Knowing this, the knight would truthfully report that the knave would say "no," while the knave would falsely report that the knight would say "no." Either way, a "no" answer indicates that you are pointing to the door to heaven.

The other case proceeds along similar lines. If you are pointing to the door to hell, the knight would say "no" when asked whether it is the door to heaven, and the knave would say "yes." Knowing this, the knight will truthfully report that the knave would say "yes," while the knave would falsely report that the knight would say "yes." Either way, a "yes" answer means you are pointing to the door to hell.

Thus, on hearing the answer to your question, you do the opposite of what you are told. □

This is the standard solution. Our clever question effectively tricks the guards into identifying the correct door. An alternative approach is to trick the guards into revealing who is guarding which door. A question that does this is: "Is the knight guarding the door to heaven?"

Suppose the guard you are addressing really is guarding the door to heaven. If he is the knight, then he will truthfully answer "yes," and if he is the knave, then he will lie and answer "yes." If the guard you are addressing is not guarding the door to heaven, then the knight will truthfully answer "no," while the knave will lie and answer "no." Either way, a "yes" answer indicates that you are speaking to the guard at the door to heaven, while a "no" answer indicates that you are speaking to the guard at the door to hell.

Both of these solutions to the heaven/hell problem allow us to open the correct door, but they do not allow us to determine which guard is the knight and which is the knave. This makes sense. The puzzle's solution space consists of four possibilities: Each door can be one of two things, and each guard can be one of two things. A single binary question cannot distinguish among four possibilities.

There is a humorous, if somewhat sketchy, approach to this problem that *does* allow us to determine not only which is the correct door, but also the personality of each of the guards.

The idea is this: Suppose we point to one of the doors and ask one of the guards, "Is this the door to heaven?" For simplicity, assume throughout this discussion that you really are pointing to the door to heaven. The knight will immediately answer "yes," and the knave will immediately answer "no." This will not be helpful to you.

But now ask the more complex question, "If I asked you if this is the door to heaven, what would you say your answer would be?" Now there are two layers to consider. Let us refer to the full, complex question as Q, and refer to "Is this the door to heaven?" as the *direct question*.

The knight will answer the direct question with "yes," and then he will truthfully say that he would answer "yes" to Q. For the knave, however, things are more complex. He reasons, "I know this *is* the door to heaven, so I would answer the direct question with 'no.' But when asked what answer I would give to the direct question, I would lie and say 'yes.'" The point is that for the knight, the answer is immediate, but for the knave, a thought process is involved.

Now add another layer of nesting: "If I asked you what you'd say, if I asked you what you'd say if I asked you if this is the door to heaven, what would you say your answer would be?" The knight is unfazed. No matter how many layers of nesting there are, the knight will immediately say "yes" and that is the end of it. For the knave however, it is very difficult to parse the question. He reasons: "Since this is the door to heaven, I would say 'no' to the direct question. If asked what I would say to the direct question, I would lie and say 'yes.' So asked what I would say if asked what I would say when asked the direct question, I would lie again and say 'no.'" Surely it takes the liar some time to work through this.

The solution, then, is to put exactly 100 layers of nesting in front of the question. The knight, as always, does not care. For him, the answer to the direct question is yes, and all the nesting in the world will not sway him from that answer. But the knave will stand there in dumbfounded silence while he tries to work out all the reversals in his answers. Thus, if you get an immediate answer, then you are addressing the knight, and if you get a delayed answer, then you are addressing the knave. If you are addressing the knight, then you can simply take his answer at face value. Since there is an even number of layers of nesting, the liar will eventually reply precisely as he would to the direct question. Therefore, do the opposite of what he tells you.

11.4 The Nelson Goodman Principle

Let us now consider a tricky variant on the basic heaven/hell puzzle.

Puzzle 61. *As before, there are two doors, one leading to heaven and the other leading to hell. There are two guards in the room, but it is possible either that both are knights, both are knaves, or that one is a knight and one is a knave. What one question can we ask to determine the correct door to take?*

In Section 11.3, we found two solutions to the classic heaven/hell puzzle. If you review the reasoning underlying those solutions, you will find that both depended critically on the assumption that the two guards were of different types. Since that assumption is no longer valid, we shall have to look elsewhere for a solution.

What is needed is a question whose answer is informative regardless of the nature of the person to whom it is addressed. There is a question that works, as I shall explain momentarily. Raymond Smullyan presents the trick in several of his books, attributing its discovery to the philosopher Nelson Goodman. If you imagine that p represents any proposition whose truth value you wish to determine, there is a single question you can ask whose answer will tell you whether or not p is true.

The idea is to make use of biconditional statements. Recall that biconditional statements are true precisely when their two parts have the same truth value (either both true or both false), and false if they have different truth values.

With that in mind, ask the question: "Is p true if and only if you are a knight?" For simplicity, suppose the person we are speaking to is named Jane. We have the following cases to consider:

- Suppose p is true. If Jane is a knight, then both parts of the biconditional are true. Therefore, the correct answer is yes, and that is what Jane will say. If Jane is a knave, then the two parts of the biconditional have different truth values. It follows that the correct answer

TABLE 11.1.
The reasoning underlying the Nelson Goodman principle.

Value of p	Jane Is a ...	$p \leftrightarrow k$	Correct Reply	Jane Says
True	Knight	True	Yes	Yes
	Knave	False	No	Yes
False	Knight	False	No	No
	Knave	True	Yes	No

Note: p denotes a proposition whose truth value we seek, and k denotes the proposition that Jane is a knight.

is no, but Jane will lie and say "yes." So, if p is true, then Jane will say "yes," regardless of whether she is a knight or a knave.

- Suppose p is false. In this case, the analysis is essentially the reverse of what we just saw. If Jane is a knight, then the correct answer is no, and that is what Jane will say. If Jane is a knave, then the correct answer is yes, but she will lie and say "no." So, if p is false, then Jane will say "no," regardless of whether she is a knight or a knave.

To summarize, a "yes" answer implies that p is true, while a "no" answer implies that p is false. This reasoning is summarized in Table 11.1. We can now solve Puzzle 61.

Solution: Armed with the Nelson Goodman principle, we can devise a third solution to the heaven/hell puzzle that works for the variant at the start of this section. Point to either door and ask either guard: "Is this the door to heaven if and only if you are a knight?" A yes answer indicates that you are pointing to the door to heaven, while a no answer indicates that you are pointing to the door to hell. Unlike our previous solutions, this one makes no assumption regarding the distribution of knights and knaves among the guards. □

Let us close by noting that there is an alternative phrasing for the Nelson Goodman principle. We could ask: "Are you the type who could claim that p is true?" This phrasing seems more natural, but the analysis is very similar to that for the biconditional version. Suppose p is true. Then the type who could claim that p is true is a knight. Thus, a knight will truthfully reply "yes," while a knave will falsely reply "yes." Now suppose that p is false. In this case, the type who could claim that p is true is a knave. Thus, a knight would truthfully reply "no," while a knave would falsely reply "no." Either way, a "yes" answer implies that p is true, while a "no" answer implies that p is false.

11.5 Generalized Nelson Goodman Principles

If we find ourselves on a standard knight/knave island, the Nelson Goodman principle allows us to learn the truth value of an arbitrary proposition. However, we have also considered islands with more diverse casts of characters, and we might wonder whether there are extensions of the principle applicable to them. For example, suppose we are on a knight/knave island on which some of the inhabitants are sane and some are insane. Is there a single question we can ask an inhabitant of such an island to learn the truth of an arbitrary proposition?

Indeed there is! In fact, we need only a small variation on the original form of the principle. Imagine you are on the island and speaking to someone named Joe. Let us say that Joe is *reliable* if he is the sort of person who makes true statements, and *unreliable* if he makes false statements. Then we can ask simply, "Are you reliable if and only if p?" Having read to this point, you are probably proficient at working through the cases for such puzzles, but let us spell them out anyway. Note that it is sane knights and insane knaves who are reliable, and insane knights and sane knaves who are unreliable.

- Suppose p is true and Joe is reliable. Then the biconditional is true. If Joe is a sane knight, then he will believe the biconditional is true and truthfully answer "yes." If Joe is an insane knave, then he will believe the biconditional is false, but then answer falsely by saying "yes."
- Suppose p is true and Joe is not reliable. Then the biconditional is false. If Joe is an insane knight, then he will believe the biconditional is true, and he will say "yes," believing he is answering truthfully. If Joe is a sane knave, then he will believe the biconditional is false, but will falsely answer "yes."
- Suppose p is false and Joe is reliable. Then the biconditional is false. A sane knight will believe it is false and truthfully answer "no." An insane knave will believe the biconditional is true, but then lie about his belief by answering "no."
- Finally, suppose p is false and Joe is unreliable. In this case, both parts of the biconditional are false, making the whole statement true. If Joe is an insane knight, then he will believe the biconditional is false, and he will say "no," believing he is answering truthfully. If Joe is a sane knave, then he will believe the biconditional is true, but he will then falsely answer "no."

Incredibly, a "yes" answer indicates that p is true, while a "no" answer indicates that p is false. However, the feeling of surprise diminishes once you realize that this situation is not significantly different from the scenario considered in Section 11.4. Have another look at Table 11.1. Were we to produce a similar table for our generalized principle, it is only the second column that would change (along with changing "Jane" to "Joe" in the final column). Instead of distinguishing the cases where Jane is a knight or a knave, we now wish to distinguish the cases where Joe is reliable or not reliable. Otherwise, everything goes through unchanged.

Once this is understood, the Nelson Goodman principle can be adapted to handle very general situations. Let us consider an especially complex island. On this island, the following facts are true:

1. Every inhabitant is a knight or a knave.
2. Male knights tell the truth, while male knaves lie. For women it is the reverse: Female knights lie, while female knaves tell the truth.
3. All of the inhabitants are either sane or insane. Sane people only hold true beliefs, while insane people only hold false beliefs.
4. When you ask an inhabitant a yes/no question, he answers by showing you either a red card or a black card. One color means yes while the other means no, but you do not know which is which.
5. Moreover, the inhabitants are inconsistent in their understanding of the colors. Some use red and black to mean yes and no, respectively, while others reverse this.

Here is your puzzle, adapted from Smullyan (2009, 37–40):

Puzzle 62. *Devise a Nelson Goodman principle for this island. That is, if p is an arbitrary proposition, devise a question whose answer will establish the truth or falsity of p, regardless of to whom the question is addressed.*

This truly sounds hopeless, but it can be done! In fact, Smullyan's solution to this problem is incredibly general.

Solution: The key is that a given native's internal thought process is not what is important. Ultimately, we do not care whether the speaker is truthful or not, sane or insane, male or female, or what the colors on the cards mean to him or her. We only care about the truth or falsity of p.

Now, let us imagine that on our island, each native responds to a yes/no question in one of two ways, which we shall call *Response 1* and *Response 2*. Depending on the island, these responses might be the acts of saying "yes" or "no," or flashing a red card or a black card, or answering in a foreign language with some word that means yes or some word that means no.

TABLE 11.2.
The characteristics of Type 1 and Type 2 natives.

Native's Type	Correct Answer to Q	Native's Reply to Q
Type 1	Yes	Response 1
	No	Response 2
Type 2	Yes	Response 2
	No	Response 1

Note: Q denotes an arbitrary yes/no question.

Let us now make two further definitions. A native is said to be of *Type 1* if she gives Response 1 to yes/no questions whose correct answer is yes and Response 2 to yes/no questions whose correct answer is no. A native is of *Type 2* if her behavior is the reverse of Type 1. A Type 2 native gives Response 2 in reply to yes/no questions whose correct answer is yes, and Response 1 to yes/no questions whose correct answer is no. These definitions are summarized in Table 11.2.

Given these definitions, what happens if we ask a native: "Are you of Type 1 if and only if p is true?" Notice that the correct answer to this question is yes precisely when both parts of the biconditional have the same truth value. That is, the correct answer is yes only when:

(1) The speaker is of Type 1, and p is true, or
(2) The speaker is of Type 2, and p is false.

There are four cases to consider: The native could be Type 1 or Type 2, and he could give Response 1 or Response 2.

First suppose the native gives Response 1:

- If the native is of Type 1, then Response 1 indicates that the correct answer to the question is yes. It follows that either (1) or (2) holds. Since the speaker is Type 1, we know that (2) does not hold. So (1) holds, and we conclude that p is true.
- If the native is of Type 2, then Response 1 indicates that the correct answer to the question is no. For this to be the case, it must be that (2) does not hold. (Of course, (1) also does not hold, but that is not relevant to our analysis.) Since (2) does not hold but the native really is of Type 2, we conclude that it is not the case that p is false. In other words, p is true.

Thus, Response 1 indicates that p is true. If instead the native gives Response 2, then the cases work out like this:

- If the native is of Type 1, then Response 2 indicates that the correct answer to the question is no. Then neither of our two options holds, and in particular, (1) does not hold. Since the native really is of Type 1, we conclude that p is actually false.
- If the native is of Type 2, then Response 2 indicates that the correct answer is yes. Since the native is of Type 2, it must be that (2) is true. This again implies that p is false.

Thus, Response 2 indicates that p is false. To summarize, Response 1 tells you that p is true, and Response 2 tells you that p is false. This holds regardless of what is going on in a native's mind. □

It really seems like a magic trick, does it not?

11.6 Coercive Logic

The solutions to the puzzles in this section are presented at the end of the chapter.

Suppose we are playing a game in which I get to ask you a question, and you are obligated to answer truthfully. How would you handle this question: "Will you either answer no to this question or pay me one hundred dollars?"

What could you do? My question is equivalent to asking whether you will do at least one of the following two things:

1. You will answer no.
2. You will pay me one hundred dollars.

If you answer no, then you will have made item 1 true, meaning that you did not answer truthfully. It follows that you must answer yes. But for this to be a truthful answer, one of the two items must be true. Since item 1 is not true, we must have that item 2 *is* true. So you had better pay me a hundred dollars, on pain of not holding up your end of the deal.

Smullyan refers to this sort of thing as *coercive logic*. That is, the logic of the situation compels you to a certain behavior, even if you regard that behavior as undesirable. The next two puzzles further illustrate the idea. Note that Smullyan embedded these puzzles in stories based on characters from the tales of the Arabian nights.

Puzzle 63. *"Here is one you might like," said Scheherazade. "A certain sultan had two lovely daughters and said to a visiting prince, 'You seem like a personable young man, and I like you, so here is what I am going to do: You are to make a statement. If the statement is true, then I will give you my younger daughter in marriage, but if the statement is false, then you may not marry her.'"*

"Now, it so happened that the prince had his heart set on the elder daughter, not the younger! He then cleverly made a statement that forced the sultan to give him the elder daughter. What statement would do this?" (Smullyan 1997, 85–86)

Puzzle 64. *"There is another version," said Scheherazade. "According to this one, the prince was greedy and wanted to marry both daughters. What statement could the prince make to force the sultan to give him both daughters?"* (Smullyan 1997, 86)

After several more puzzles of this general sort, Smullyan builds up to what he refers to as the ultimate in coercive logic. Note that some of the ideas underlying this puzzle will recur in Chapter 14.

Puzzle 65. *Suppose I offer you a million dollars to answer either yes or no, and you have the option of answering either truthfully or falsely— you may either tell the truth or lie at your pleasure. Surely now you are completely safe, are you not? How can mere words coerce you into paying me money when you don't even have to be truthful in your answer? Why can't you simply answer yes or no at random and then refuse to pay me anything? How much safer can you be?* (Smullyan 1997, 101–102)

11.7 Smullyan as a Writer

With this section, we come to the end of that portion of the book devoted exclusively to Raymond Smullyan's puzzles. There is one more aspect of his work we should consider, before moving on to the next topic.

In our discussion of Lewis Carroll's puzzles, we noted the exceptional creativity of his presentation. Any pedantic textbook author might challenge a student to show that "Some As are Bs" and "No Bs are Cs" together imply that "Some As are not Cs." But it takes Carroll's creativity to think instead of asking the student to show that "Some oysters are silent" and "No silent creatures are amusing" imply that "Some oysters are not amusing." Carroll's version is not only more fun to read but also more pedagogically effective.

Similar consideration should be given to Smullyan's virtues as a writer. If you focus solely on his ingenuity in devising puzzles, then you will miss an important aspect of his work. Like Carroll, Smullyan was highly inventive in his manner of presentation. The idea of embedding his puzzles in stories based on famous characters from history and literature made them very engaging and enjoyable.

An especially poignant example of this technique is found in Smullyan's book *King Arthur in Search of His Dog* (Smullyan 2010). When King

Arthur's dog goes missing, the king embarks on an epic quest to find him. This inevitably leads to various characters making cryptic statements about the dog's possible whereabouts. I recall getting emotional when, after many pages of this, it seemed that all leads had been exhausted with no sign of the dog. (Sometimes I get emotional over more traditional textbooks, but for different reasons.)

In an essay about analogies between mathematical exposition and fictional narratives, mathematician Timothy Gowers writes:

> If the novel is describing a series of events, then those events will have a natural order, but there is much more to a good novel than "A happened, then B happened, then C happened, ...," and the order in which information is revealed to the reader is often not chronological. Likewise, in a mathematical presentation, statements come in a logical order, ... and this order frequently differs both from the order in which the statements are discovered and from the order in which they should be presented if they are to be understood most easily. The result of all this is that mathematicians who wish to communicate their ideas effectively face many of the same challenges as novelists. (Gowers 2012, 211)

Gowers goes on to identify one particular attribute that a narrative, either mathematical or fictional, has to one degree or another. He refers to this attribute as "vividness," and I am sure you can surmise what he has in mind.

This notion perfectly captures what sets Carroll and Smullyan apart from the average mathematical expositor. Their use of vividness, and their understanding that there is more to good mathematical writing than concision and rigor, is a reminder of what made you like mathematical puzzles in the first place. Too few mathematicians manage to grasp this point.

11.8 Solutions

Puzzle 63. The prince could say: "You will not give me either daughter in marriage." If we decide this statement is true, then the sultan is required to give his younger daughter in marriage, which immediately falsifies the statement. So, we must assess this statement as false. For the statement to be false, the prince must be given one of the daughters in marriage. But the rules do not permit the younger daughter to be given for a false statement. It follows that the prince must be permitted to marry the elder daughter.

Puzzle 64. This time the prince could say: "You will not give me the younger daughter in marriage unless you also give me the older daughter."

If we assess this statement as false, then the sultan must give his younger daughter in marriage, but not the older daughter. But this is impossible, since the younger daughter cannot be given for a false statement. So we must assess the statement as true. The rules say that for a true statement the younger daughter must be given, but since the statement is true, this implies the older daughter must be given as well.

Puzzle 65. The key to the puzzle is that the respondent is not quite allowed to answer randomly. The answer must be either true or false. With that in mind, suppose we ask: "Will you either truthfully answer no to this question, or falsely answer yes, or pay me two million dollars?"

We are asking whether one of the following three alternatives holds:

1. You will truthfully answer no.
2. You will falsely answer yes.
3. You will pay me two million dollars.

You cannot falsely answer yes, since then the second alternative would hold, leading to a contradiction. Likewise, you cannot truthfully answer no, since then the first alternative holds, another contradiction.

Now suppose you truthfully answer yes. Then one of the three alternatives must hold. But it is quickly seen that only the third can hold, forcing you to pay me two million dollars. What if you falsely answer no? A truthful "no" answer would entail that none of the three alternatives holds. Therefore, a false "no" answer entails that at least one holds. Like before, the only one that can hold is that you will pay me two million dollars.

Either way, you will pay me two million dollars.

PART IV

Puzzles Based on Nonclassical Logics

CHAPTER 12

Should "Logics" Be a Word?

Such was Aristotle's dominance over logic for most of its history that it was hard not to notice a certain triumphalism among his followers. Immanuel Kant expressed a common view when he wrote, in 1787 (comparing the development of logic to the "sure path" traveled by successful sciences):

> That logic has already, from the earliest times, proceeded upon this sure path is evidenced by the fact that since Aristotle it has not [been] required to retrace a single step. . . . It is remarkable also that to the present day this logic has not been able to advance a single step, and is thus to all appearance a closed and completed body of doctrine. (Kant 2003, 17)

As we saw in the brief history presented in Chapter 8, Aristotle's dominance over logic did not seriously wane until the development of mathematical logic in the nineteenth and early twentieth centuries. This new logic quickly assumed a dominance of its own, which led to its own expressions of triumphalism. For example, here is Bertrand Russell commenting on the *Organon* as a whole:

> I conclude that the Aristotelian doctrines with which we have been concerned in this chapter are wholly false, with the exception of the formal theory of the syllogism, which is unimportant. Any person in the present day who wishes to learn logic will be wasting his time if he reads Aristotle or any of his disciples. . . . Throughout modern times, practically every advance in science, in logic, or in philosophy has had to be made in the teeth of the opposition from Aristotle's disciples. (Russell 1972, 202)

You have probably guessed what is coming. The notion that classical logic is the one uniquely correct logic is not as widely held as it once was. In fact, today it is a lively source of philosophical debate whether it even makes sense to speak of a uniquely correct system of logic. Perhaps we should just talk about different systems of logic, each useful and correct in its own domain. Nowadays many systems of logic are on offer, each with its advocates and defenders.

That is why people sometimes talk about "logics," plural. Let us take a brief look at a few of the issues.

12.1 Logical Pluralism?

Classical logic has two big things going for it. The first is that its fundamental notions seem so natural and intuitive that you might think any system of logic just has to be based on them. The second is that it has paid enormous dividends for mathematics. Let us consider each of these points for a moment.

As we have previously discussed, at the heart of classical logic are two basic ideas. The first is the *principle of bivalence*. This is the principle that says there are only two truth values: true and false. There is no third value. We do not have the option of assessing a proposition to be neutral, or of saying that it is true in one sense but false in another. In normal discourse, we often do avail ourselves of such options, but such judgments should play no role in formal reasoning. Closely related to the principle of bivalence is the *law of the excluded middle*, which asserts that for any proposition, either that proposition or its negation is true.

The second basic idea is the *law of noncontradiction*, which asserts that a proposition cannot be simultaneously true and false. Equivalently, we can say that a proposition and its negation cannot both be true. When combined with the law of the excluded middle, we derive the principle that for any proposition, one among that proposition and its negation is true, and the other is false.

Now, seriously, what could be more commonsensical than that? Any assertion you make either does, or does not, correspond to the facts (rival theories of truth notwithstanding). How could it do both simultaneously? It is gibberish even to assert the possibility. And while it is certainly true that everyday conversation confronts us with countless assertions too vague or ambiguous to be assigned a definite truth value, such statements should simply be exiled to the phantom zone where deductive logic does not apply. Sharpen up your propositions, and then we will talk about logic!

Moving on to the second point, recall that classical logic was developed specifically as a tool for mathematics. For this purpose, it has been enormously successful. So successful, in fact, that mathematicians pay no attention at all to what they see as so much philosophical prattle about the possibilities for nonclassical logic. We use classical logic, mathematics works, and no instance has ever been adduced of a mathematical assertion that was classically proved later turning out to be incorrect. To anyone who suggests we are doing it wrong, we respond with profanity.

These two points amount to a strong case that classical logic is really on to something. It accords well with common sense, and it produces good results.

Still, classical logic is vulnerable to three general sorts of challenges. The first is that by ignoring problems of vagueness, we are ensuring that deductive logic will be applicable only in the rarefied worlds of mathematics and formal systems. If our system of logic can only handle absolute precision, then it is not useful in most aspects of our lives. In this view, insisting on a sharp dichotomy between true and false is too restrictive. This sort of consideration has led people to study systems of logic in which the principle of bivalence does not hold. Such systems are typically referred to as *many-valued logics* (when the number of truth values is finite) or as *fuzzy logic* (when infinitely many truth values are permitted).

The second challenge is relatively mild: There seem to be concepts relevant to good reasoning to which classical logic pays no attention. For example, it is not part of classical logic that a proposition could be true at one time but false at another time. (It is sometimes said that classical logic is "tenseless.") Nor does classical logic make a distinction between statements that are necessarily true, as opposed to those that are only possibly or contingently true. Another deficiency arises from moral reasoning, since classical logic has no concept of what we "ought" to do, or of what we are "permitted" to do.

These challenges have led people to develop extensions of classical logic. Systems of logic cognizant of notions of time are called *temporal logics*. Systems that make distinctions among the different modes in which a statement can be true (such as necessarily true versus possibly true) are referred to as *modal logics*. Systems in which "ought" and "permitted" are meaningful are called *deontic logics* (from a Greek word meaning "that which is binding"). None of these systems represent a rejection of classical logic, in the sense that any argument form that is classically valid is still accepted as valid under these other systems. Instead, these systems enhance classical logic by adding other sorts of notions and connectives to it.

The third challenge is more serious. It comes from those who insist that classical logic is, in some way, just flat wrong as a description of proper reasoning. You see, classical logic does not *always* accord with common sense.

Classical logic employs the material conditional, which treats the conditional statement $P \rightarrow Q$ as equivalent to $\neg P \vee Q$. This is fine for mathematicians, since when we say "If P, then Q," we really are saying nothing more than, "Either P is false or Q is true." (We could probably avoid a lot of confusion by not using if–then at all and instead expressing everything in the form of disjunctions.) In everyday discourse, however, conditional statements are, in general, asserting more than a mere disjunction. When someone asserts, "If P, then Q" in normal conversation, it is likely that she is asserting that P is in some way the cause of Q. At a minimum, she is asserting that P and Q are relevant to one another. But classical logic has no concept of causation or relevance.

For example, classical logic says that "If $2 + 2 = 5$, then Santa Claus exists" is true, and so is "If Santa Claus exists, then $2 + 2 = 4$." Intuition is not amused.

Here is another quirk of classical logic: It turns out that a contradiction implies any statement at all. To see why this is so, let P and Q represent any two propositions. Let us take as a premise the contradiction $P \wedge \neg P$. Then we can reason like this:

$P \wedge \neg P$.

Therefore, P.

So, we can assert $P \vee Q$.

But, $\neg P$.

Therefore, Q.

The second line follows from the first, since the parts of a conjunction must be true individually. The third line follows from the fact that a disjunction is true as soon as any one of its parts is true. The final line follows from the two before it by basic properties of disjunctions. We have just derived an arbitrary proposition from an arbitrary contradiction.

For example, classical logic says that the following is a valid argument: "The number 2 is both prime and not prime; therefore Santa Claus exists." This seems strange. The principle that anything follows from a contradiction is known by the Latin phrase *ex falso quodlibet*, or more simply as the *principle of explosion*.

Such considerations might lead one to think that classical logic may not be the last word after all. Systems of logic that try to impose a relevance relation between the parts of a conditional statement are called *relevance logics*. Systems in which the principle of explosion does not hold are called *paraconsistent logics*.

At this point, you might be thinking that Aristotle, and his followers, would be aghast. We could imagine them saying: "What nonsense is this? What sort of insanity leads one to speak of different logics, plural? There is only one sort of correct reasoning, and logicians are tasked with the job of working out what it is."

Through most of the history of logic, most philosophers would generally have agreed. Nowadays the matter seems more unsettled, and the question of "logical pluralism" is one that is treated with great seriousness. Philosopher Susan Haack helpfully summarizes the discussion like this:

> It will be useful to start by distinguishing, in a rough and ready way, three broad kinds of response to the question whether there is a uniquely correct logical system:
>
> *monism:* There is just one correct system of logic.
>
> *pluralism:* There is more than one correct system of logic.
>
> *instrumentalism:* There is no 'correct' logic; the notion of correctness is inappropriate. (Haack 1978, 221)

It is very mysterious, however, what "correct" means in this context. Haack provides the following elaboration:

> [R]oughly, a logical system is correct if the formal arguments which are valid in that system correspond to informal arguments which are valid in the extra-systematic sense, and the [well-formed formulas] which are logically true in the system correspond to statements which are logically true in the extra-systematic sense. The monist holds that there is a unique logical system which is correct in this sense, the pluralist that there are several. (Haack 1978, 222)

The idea seems to be that we come to our study of logic with certain intuitive understandings of what is valid, and that is the standard we use to assess the correctness of a system. In a more recent discussion, philosopher Roy T. Cook gives a very similar account of what is meant by "correct:"

> [A] logic . . . is correct if and only if, for any way of interpreting the non-logical vocabulary, the logic validates a particular argument if and only if the natural language statement corresponding to the conclusion of that argument is a logical consequence of the natural language statements corresponding to the premises of that argument. (Cook 2010, 495–496)

Cook seems to take for granted that, when working with natural languages, we know validity when we see it. However, it is very unclear whether such intuitions as we have regarding logical validity can really bear the burdens Haack and Cook are placing on them.

Leaving that aside for the moment, there is no question that you can devise formal systems that differ from the one used in classical logic, and, if it amuses you to do so, you can then investigate the properties of those systems. By all means make up whatever definitions you like, and then feel free to prove whatever theorems follow from your setup. I certainly will not try to stop you. If you want to countenance more than two truth values, or if you want to define connectives and rules of inference other than those employed by classical logic, there is no philosophy police to tell you not to. This is not what people mean when they ask whether there is more than one correct logic.

Instead, the issue has to do with the notion of logical consequence. First and foremost, logic is the study of what follows from what. The standard approach to this problem is to say that certain arguments are valid because they are specific instances of abstract forms that are themselves deemed to be valid. In speaking of "abstract forms," the whole point is that the validity of the argument in no way depends on the subject matter of the propositions involved. Logic is meant to discover universal principles of correct reasoning. In the traditional view, it does not make sense to say that an argument form might be valid in one context but invalid in another. The modern monists defend this view, while the modern pluralists demur.

12.2   Is Classical Logic Correct?

The existence of this debate reveals a cultural difference between philosophers and mathematicians. That is, between those who study logic for its own sake and those who merely use logic as a tool in their work. Mathematicians regard it as obvious that classical logic is the correct logic, at least when applied to mathematical problems. If intuition balks at some of its consequences for nonmathematical discourse, then so much the worse for intuition. Classical logic improves our lives every day through the myriad successes of applied mathematics. Are we really to give that up because the principle of explosion seems a little weird?

Notice, though, that I did not say classical logic is correct, full stop. I restricted it to a particular domain of discourse. Classical logic is obviously correct *for mathematics*. We might yet decide that, just as Aristotelian logic was absorbed into propositional logic, and propositional logic was absorbed into classical logic, so, too, might classical logic be absorbed into something else.

Alternatively, we might decide that the relationship of classical logic to the hypothetical "one true logic" is like the relationship of Newtonian mechanics to Einsteinian relativity. There is a very concrete sense in which Einstein's physics is superior to Newton's as a model for the universe generally, but for nearly all practical problems, you will not go wrong using Newton's approach. Likewise, it is possible that the one true logic is effectively indistinguishable from classical logic when applied to mathematics, but is nonetheless superior as a global account of human discourse.

The physics analogy is apt in another way: It points to a difficulty with the whole debate over which is the correct logic. There is an objective standard for assessing the merits of rival scientific models: Which one is better able to predict the results of experiments? When applied to normal problems, the predictions made by Newton and Einstein are so similar that no measuring device can detect the difference. But when applied to certain extreme problems that lie well outside our daily experiences, Einstein does better. In contrast, there does not seem to be a clear standard for correctness in the philosophical debate over rival logics.

This is made clear by the attempts to define "correct" made by Haack and Cook in Section 12.1. Both in some way put human intuition at the center of their proposed standards for correctness. Haack referred to arguments being valid in the "extra-systematic" sense, meaning the intuitive sense of validity we use to guide our system construction in the first place. Cook simply took for granted that we recognize validity among natural language sentences when we see it.

The problem is that there really does not seem to be anything in logic so intuitively obvious that it is universally accepted. A school of thought in mathematics argues that the law of the excluded middle is not valid, and that its widespread acceptance arises from fundamentally flawed philosophical views. Ironically, this school of thought is known as "intuitionism." The law of noncontradiction is likewise challenged. Some philosophers argue that we are forced, by certain paradoxes that arise in classical logic, to countenance the reality of true contradictions. (We will revisit this topic in Section 16.4.) The technical minutiae of why someone might hold these views is well beyond what we can discuss here. What matters for us is that although these viewpoints reside on the fringe of scholarly discourse generally, they are nonetheless defensible and are in fact defended by very capable people.

Modern logical scholarship presents us with a riot of formal systems and philosophical perspectives, to the point where efforts to find something that is uniquely correct seem rather forlorn. It is unclear even what the standard for correctness should be. For these reasons, I align myself with the instrumentalists in Haack's trichotomy from Section 12.1. More specifically, I am attracted to a view known as *pragmatic instrumentalism*. Logical systems should be assessed as useful or not useful, and not as correct or incorrect in any absolute sense.

Philosopher C. I. Lewis ably expresses the basic idea:

> Any current or accepted canon of inference must be pragmatically deter-
> mined. That one such system should be thus accepted does not imply that
> the alternative systems are false: it does imply that they are—or would be
> thought to be—relatively poorer instruments for the conduct and testing of
> our ordinary inferences. (Lewis 1932, 484)

Philosopher Nicholas Rescher elaborates:

> Admitting a plurality of logics, [the pragmatic view] regards the choice
> between them not as arbitrary and wide open, but as narrowly restricted
> by a whole host of *functional* or purpose-relative considerations of effec-
> tiveness, economy, and efficiency given the inferential tasks and purposes
> upon which the logical apparatus at issue is to be deployed. (Rescher 1969,
> 221–222)

Recall that this issue arose in Section 6.4, when we discussed Lewis
Carroll's story of Achilles and the Tortoise. The Tortoise refused to accept
the conclusion of a logically valid geometrical argument and demanded
that Achilles somehow compel him to do so. Achilles attempted, unsuc-
cessfully, to provide the desired compulsion by adding premises to the
argument. I suggested that a more profitable response to the Tortoise was
to argue on instrumentalist grounds. Logic is not about forcing anyone to
do anything, but if the Tortoise wants to get good results in his geometrical
studies, then he would do well to heed what the logicians are saying.

My objection to noninstrumentalist views of logic is similar to the
objection I raised with regard to the discourse on Achilles and the Tortoise. We are introducing vague terms into the discussion—"correct" and
"incorrect," in this case—for no good purpose.

This issue is difficult, and it has given rise to the inevitable vast literature. Perhaps I am wrong to be attracted to instrumentalism, but the view
has a definite practical benefit. When I am confronted with people advo-
cating for outré ideas in logic, I do not have to argue with them. I can
just smile and say: "Well, I am getting good mathematical results with
classical logic, so I am just going to keep on using it. I am sure that your
preferred system of nonclassical logic is adorable, but come back when
you can actually do something with it."

12.3 Applications of Nonclassical Logic

Here's the thing, though. Increasingly, people *are* finding applications for
nonclassical logic. Just as the needs of mathematicians required a shift

from Aristotelian to classical logic, so do the needs of computer scientists sometimes require a shift from classical to nonclassical logic.

Readers of a certain age will remember the 1960s American television series *Star Trek*, in which the intrepid explorers of the starship Enterprise, led by Captain James T. Kirk, explored the cosmos and encountered alien races. A recurring theme in the show was the conflict between human morality and emotion on the one hand, and cold impersonal logic on the other.

In an especially memorable episode, the crew of the Enterprise is imprisoned on a planet ruled by powerful, but ultimately benevolent, androids. The androids believe that humans need their help to avoid self-destruction, and they plan to take the Enterprise so that they may offer their unwanted help to other planets as well. The prospects for escape seem grim until the crew discovers that the androids become confused in the face of inconsistent data. They proceed to confront the androids with so many verbal contradictions and instances of irrational behavior that one by one, the androids burn out their circuits and shut down. In the episode's climax, Captain Kirk and another character, named Mudd, confront Norman, the leader of the androids. Kirk tells Norman that everything Mudd says is a lie. Mudd then tells Norman, "I am lying right now." Norman burns himself out trying to resolve the paradox, and the crew is able to escape.

Those silly androids were using classical logic, you see. Had they been using a paraconsistent logic, the Enterprise crew might be imprisoned on that planet to this day.

Something surprisingly similar to the *Star Trek* scenario arises in the modern science of information processing. It is often said that we are living in the information age, and computers are increasingly expected to handle, and draw inferences from, very large datasets. It is nearly inevitable that there will be inconsistencies in the data. Unless we want all of our databases to suffer the fate of the *Star Trek* androids, we might want to investigate the possibilities for employing paraconsistent logics in these contexts. Interestingly, the branch of artificial intelligence tasked with programming computers to draw sensible conclusions from large datasets is known as "automated reasoning."

A similar problem arises in the construction of so-called "expert systems," by which we mean computers programmed to simulate the abilities of human experts. Such systems are typically endowed with some base of facts as a starting point and are then programmed with certain inference rules from which to derive novel conclusions from the facts. However, when there is the danger of vagueness in the data, it might make sense to employ a many-valued or fuzzy logic. This is an active area of research in the fields of artificial intelligence and theoretical computer science.

Nor is computer science the only area in which applications of non-classical logics are being actively investigated. To date, they have found applications in philosophy (particularly metaphysics), mathematics, and even in linguistics (Gottwald 2015).

You might wonder, though, whether such applications that have been found really have any relevance to the debate over logical pluralism mentioned earlier. Does the fact that it is sometimes useful to program computers to handle data in ways that differ from classical logic show that there is more than one correct account of logical validity (whatever *that* means)?

See Section 12.4 to find some references on *that* little poser. For now I simply note that whatever their status among rival camps of philosophers, nonclassical logics are definitely here to say. Logic has gone from the rigidity of the Aristotelian perspective (which held that the entirety of human reasoning could be captured in the very simple mechanics of categorical syllogisms) to the remarkably permissive modern attitude, in which it seems like any old invented-in-the-shower formal system or fever dream of a philosophical perspective is vigorously discussed and debated.

Regardless of their potential applications for the betterment of humanity, however, these modern logics certainly provide plenty of fodder for empuzzlers.

12.4 *Notes and Further Reading*

The subject of instrumentalism also arises in science, where a vigorous philosophical debate is proceeding between those who take a realist view of the discipline versus those who take an antirealist view. The problem is this: Many scientific theories posit the existence of unobservable entities like subatomic particles. These theories seem to work well, in the sense that they accurately predict the results of experiments. The realist view is that the agreement between theory and experiment arises because the unobservable entities actually exist out there in the real world, and our theories are an accurate description of how they actually behave. The antirealist in some way denies this. Instrumentalism, in this context, is a sort of mild antirealism. It holds that the goal of science is not capital T truth, but instead, the production of models and theories that render nature predictable and controllable. Those unobservable entities might be real and they might just be useful fictions, but in their professional roles scientists just ought not to care one way or the other about this. Okasha (2002, 58–76), and the references contained therein, is a gentle way into this discussion.

The Polish logician Jan Łukasiewicz introduced the first fully worked-out system of three-valued logic in 1917. His intended applications were to modal logic, the idea being that the third truth value would represent "possibly true," but this idea turned out not to be so useful. In its early days, applications to handling problems of vagueness were also on the agenda. By the 1960s, when computer science was still in its infancy, the subject had reached a level of sophistication that merited a book-length treatment (Rescher 1969). For a sometimes skeptical look at the potential value of nonclassical logics, see the book by Haack (1974).

Of course, both of those books long predate the modern applications of nonclassical logics in computer science. For more up-to-date discussions regarding the most important nonclassical systems, consult the books by Beall and Van Fraassen (2003), Priest (2008b), and Burgess (2009). On the subject of logical pluralism generally, the papers by Cook (2010) and Eklund (2012) are good places to start. For a full-throated defense of logical pluralism, try the book by Beall and Restall (2005). The paper by Field (2009) offers reasons for skepticism. A recent defense of logical instrumentalism is given by Kissel (2016).

I mentioned that even something as fundamental, and as seemingly unassailable, as the law of noncontradiction is today open to debate. In Section 1.2, I mentioned dialetheism, which, recall, is the idea that some contradictions are true. (We will visit this idea again in Section 16.4.) For a thorough examination of the issues on both sides of the law of noncontradiction, see the anthology edited by Priest, Beall, and Armour-Garb (2004). The papers in this book are difficult and technical, but I did get a chuckle out of the following statement, from the chapter authored by Stewart Shapiro, who is critical of dialetheism:

> The dialectic between dialetheism and its critics is not straightforward. It is difficult for the critic to know what to say. One can refute other opponents by showing that their views lead to contradiction, especially if the inconsistency comes by the opponent's own lights. The debate usually turns on whether the view does in fact entail a contradiction. This cuts no ice against the dialetheist, since she embraces contradictions. For her, a *reductio ad contradictionem* is not a *reductio ad absurdum*. (Shapiro 2004, 337)

CHAPTER 13

Many-Valued Knights and Knaves

Nonclassical logic certainly takes some getting used to. As noted in Chapter 12, one of the selling points of classical logic is its naturalness, in the sense that it captures many of our intuitive notions of how logic is supposed to be. Certainly if you go looking for trouble, you can find it in the form of the paradoxes of material implication or the principle of explosion, but those are things only philosophers worry about. The simple fact is that there is no scenario in your daily routine where you will go wrong by applying classical logic, which makes the whole idea of nonclassical logic seem unimportant.

We have seen, however, that there *are* applications of such logics, suggesting that we might want to spend some time engaging with them. And since this book is all about using puzzles as a gateway for thinking about logic, the present chapter is devoted to the following question: What would it look like if the knights and knaves employed a many-valued system of logic?

Many of the puzzles in this chapter have been previously published in Rosenhouse (2014, 2016).

13.1 The Transitional Phase

To review, classical logic has certain distinctive features. It employs precisely two truth values, typically referred to as "True" and "False." This is known as the *principle of bivalence*. Two other standard principles are the *law of noncontradiction* and the *law of the excluded middle*. These two principles can be viewed as saying that the two truth values are mutually exclusive and jointly exhaustive. That is, every proposition is either true or false, but not both simultaneously. A *nonclassical logic* can then be viewed roughly as any system of logic that rejects at least one of these principles.

These principles seem so natural that on hearing that someone has presumed to reject them, our first reaction might well be annoyance. This is unwarranted, since a moment's thought reveals that classical logic simply ignores many of the realities of normal conversation. A rigid dichotomy between true and false obscures the fact that many assertions are vague. Sometimes we wish to say that a statement is partly true and partly false, as when the statement, "You are a child," is directed at someone who is 13 years old. Are we really comfortable assigning a definite truth value to such a statement? Classical logic says we must.

An alternative approach rejects the principle of bivalence. Perhaps what is needed are additional truth values. These new truth values would be assigned to vague statements, which is to say those for which the classical truth values of "True" and "False" do not seem to apply. Let us see how this might play out in the world of knights and knaves.

Imagine a particular pair of knight/knave islands, on which visiting anthropologists have made a startling discovery. On these islands, knighthood and knavehood are not permanent conditions. Instead, it was found that throughout their lives, people cycled back and forth between the two classes. They would be knights for a while, then they would enter a transitional phase in which they were partly knights and partly knaves, and then they would emerge at the other end as knaves. After some time as knaves, they would once more enter the transitional phase, eventually emerging at the other end as knights. All of the islanders cycled, repeatedly and unpredictably, between the two classes.

If you meet someone in the transitional phase, and you say of that person, "He is a knight," should your statement be assessed as true or as false? Since such a person is partly knight and partly knave (just as a young teenager is in some ways a child and in some ways not), neither answer seems fully satisfactory.

At this point, you might object that since knights and knaves are defined by the type of statements they make, there can be no vagueness regarding one's status: Either you always make true (or always make false) statements or you do not, and that is all there is to it. A complete response to this objection would require a proper ethnography of the island's culture. Suffice it to say that on these islands, knighthood and knavehood are more complex than they are on classical islands. They are defined by a variety of social and cultural factors, and not just on the types of statements people made.

The logicians on these islands agreed that they were poorly served by the principle of bivalence. However, they disagreed with regard to its replacement. In Sections 13.2 and 13.3, we will consider these different responses.

13.2 The Three-Valued Island

The first island decided that a third truth value was needed. This new truth value was denoted N. It applied to vague statements and was commonly thought to stand for "Neutral" or "Neither true nor false." We shall refer to an island taking this approach to logic as "three-valued." Thus, on a three-valued island, any statement is exclusively true, false, or neutral.

Of course, not just any statement could be neutral. "True" and "false" continued to mean what they have always meant. Specifically, they meant the proposition either did, or did not, correspond to reality. The third truth value was invoked solely for vague statements. In particular, it applied to assignments of knighthood or knavehood to people in their transitional phase, or to negations of such statements. (Naturally, compound propositions formed by joining statements with the standard logical connectives could also be neutral, as we shall discuss shortly.)

There is a further twist to the story, however. Entering the transitional phase represented a disconcerting loss of identity for the island's natives. Uncertain as to whether they should behave as knights or as knaves, they chose instead only to make neutral statements. For this reason, people in the transitional phase are referred to simply as "neutrals."

Thus, there are three kinds of people native to a three-valued island: knights, who only make true statements; knaves, who only make false statements; and neutrals, who are in the transitional phase and only make neutral statements. Neutrals are seen as partly knight and partly knave. For emphasis, we shall use the expressions "fully knight" and "fully knave" when referring to people not in the transitional phase.

Table 13.1 helps clarify the situation. The column headings represent statements a person might make. The row headings represent the reality of the situation. The table entries then represent the truth values that should be assigned to the statements.

Notice that the statement "X is a neutral" always receives a classical truth value. (You should be looking down the second column.) Such a statement can be either true or false, but it cannot be neutral. A person either is, or is not, in the transitional phase. If he is, then it is simply true that he is a neutral. If he is not, then it is false that he is a neutral. There is no vagueness here. Likewise, a person not in the transitional phase is definitely either fully knight or fully knave, but not both simultaneously. Again, there is no vagueness. We only encounter vagueness

TABLE 13.1.
The proper assignments of truth values to basic statements on a three-valued island.

	"X is a knight"	"X is a neutral"	"X is a knave"
X is fully knight	T	F	F
X is a neutral	N	T	N
X is fully knave	F	F	T

Note: Column headings represent statements a person might make, while row headings represent the reality of the situation.

when we attribute knighthood or knavehood to someone in the transitional phase. In this phase, a person is in some ways like a knight and in some ways like a knave, making it impossible to declare such attributions to be definitively true or definitively false. For this reason, we assign such statements the third truth value, N. For convenience, statements receiving this truth value are referred to as neutral statements. Thus, "neutral" can refer either to a type of person or to the truth value of a particular statement. This usage will not cause confusion, however, since it is always clear from the context what is intended.

These ideas take some getting used to, as our first two puzzles demonstrate.

Puzzle 66. *One day you meet Adam and Beth, who make the following statements:*

> *Adam : Beth is a knight.*
>
> *Beth : Adam is a knave.*

What can you conclude about Adam and Beth?

Solution: Such an exchange would be impossible on a classical island, because we would reason as follows: If Adam is a knight, then his statement is true. Therefore, Beth is also a knight. So Beth's statement must be true, which implies that Adam is a knave. This is a contradiction. Now suppose that Adam is a knave. Then it is false that Beth is a knight. So she must be a knave. But then her statement is false as well, implying that Adam is a knight. This is another contradiction.

This reasoning remains valid on a three-valued island, assuming we understand "If Adam is a knight/knave" to mean "If Adam is fully knight/fully knave." But now we have a new possibility to consider. It could be that Adam is a neutral. In this case, his statement receives truth value N, which is possible only if Beth herself is in the transitional phase. In this case, her statement would also receive truth value N, which is consistent with Adam being a neutral. So, Adam and Beth are both neutrals. □

Puzzle 67. *This time you meet Carl, Dave, and Enid, who make the following statements:*

> Carl : *Dave is a knight.*
>
> Dave : *Enid is a knave.*
>
> Enid : *Carl is a neutral.*

What can you conclude about Carl, Dave, and Enid?

Solution: In pondering dialogues from three-valued islands, it is helpful to focus on statements that must have classical truth values. In particular, Enid's statement cannot be neutral, for reasons we have already discussed. Therefore, Enid is either fully knight or fully knave.

Now, if Enid is fully knight, then her statement is true. Therefore, Carl really is a neutral. That means that Carl's statement is neutral, which is only possible if Dave is himself a neutral. This is not possible, since in this scenario, Dave's statement is false. We have reached a contradiction.

Therefore, Enid is fully knave. Since this implies that Dave's statement is true, he is fully knight. Since this is precisely what Carl asserts, we conclude he is fully knight as well. So the solution is that Carl and Dave are fully knight, while Enid is fully knave. □

Over the years, the island's logicians developed rules for handling negations, disjunctions, conjunctions, and conditionals. In classical logic, the truth values of such compound statements are determined entirely by the truth values of their component parts. Three-valued logic preserves this feature of classical logic.

The truth tables for negations and conjunctions in three-valued logic are shown in Table 13.2. The rules for disjunctions and conditionals are in Table 13.3.

Notice that three-valued logic is identical to classical logic when only classical truth values are used. If you look only at the entries in the corners of the tables (or in the top and bottom rows in the table for negations), you will see the familiar classical truth tables.

The only issue, then, is how to handle neutral statements. We see that the negation of a neutral statement is still neutral. This makes sense: If

TABLE 13.2.
Truth tables for negations and conjunctions in three-valued logic.

P	¬ P		∧	T	N	F
T	F		T	T	N	F
N	N		N	N	N	F
F	T		F	F	F	F

Note: In the table for conjunctions, the entries down the rows on the far left represent the truth values for the first part of the statement, while the entries across the top represent the truth values for the second part.

TABLE 13.3.
Truth tables for disjunctions and conditionals in three-valued logic.

∨	T	N	F		→	T	N	F
T	T	T	T		T	T	N	F
N	T	N	N		N	T	T	N
F	T	N	F		F	T	T	T

Note: The entries down the rows on the far left represent the truth values for the first part of the statement, while the entries across the top represent the truth values for the second part.

the statement "Joe is a child" is vague, then the statement "Joe is not a child" should also be considered vague.

To fill in the tables for conjunctions, disjunctions, and conditionals, our guiding principle has been to preserve as much of our classical intuition as possible. For example, our intuition about conjunctions is that they should only be considered true when all of their parts are true. A conjunction is false as soon as even one of its parts is false. If we now interpret N as representing "possibly true and possibly false," then the reasoning behind these tables is clear. Consider the statement "Two plus two is four, and Joe is a child." The first part of the statement is true, but we are considering the second part to be vague. Construing vagueness as possible falsity, it makes sense to evaluate the entire statement as neutral. In contrast, the first part of the statement "Two plus two is five, and Joe is a child" is false. That means the whole statement is false, regardless of how we treat the truth value of the second part.

Similar reasoning underlies the table for disjunctions in Table 13.3. We regard such statements as true if even one part is true, and false only when every part is false.

Just as with classical logic, it is the table for conditionals (see Table 13.3) that causes the most difficulty. The basic principle is to recall the classical logical equivalence

$$(P \rightarrow Q) \equiv (\neg P \vee Q).$$

In words, to assert $P \rightarrow Q$ is to say that either P is false or Q is true. Having already completed the tables for negations and disjunctions, we could now use this equivalence to complete our table for conditionals. This seems reasonable, since we want to preserve as much of classical logic as possible. So, consider the statement: "If Joe is a child, then two plus two equals four." In classical logic, the truth of a conditional's conclusion implies the truth of the whole statement. For this reason, three-valued logic evaluates this statement as true, despite the vagueness of the first part.

However, look carefully, and you will notice that the three-valued table does not use the classical equivalence to fill in the central entry, which treats conditionals in which both the antecedent and the conclusion are neutral. The equivalence would lead us to evaluate such a statement as neutral, but our table indicates that it is true. The reason is that slavishly following the classical equivalence would bring us into conflict with another principle of classical logic, known as *the law of identity*. This law asserts that for any proposition P, the proposition $P \rightarrow P$ should be evaluated as true. The logicians on the three-valued island were of the opinion that preserving this principle is more important than preserving in all cases the classical equivalence. The statement "Joe is a child" might be vague, but the statement "If Joe is a child, then Joe is a child" seems unambiguously true. That is why they completed the table as shown.

There is, therefore, a certain arbitrariness in completing these tables. They had conflicting principles, so they simply chose the system that possessed the features they found most desirable. Other logicians might have different preferences, and they might have made different choices. As it happens, the specific system of three-valued logic employed here was introduced by the Polish logician Łukasiewiecz.

Here is a practice exercise, to ensure that you are comfortable with conditionals in three-valued logic:

Puzzle 68. *You come to a fork in the road. You know that exactly one of the paths leads you to the city. A man named Fred is standing nearby, and you ask him which path to take. He replies:*

> Fred: *If I am a knight, then the left fork will take you to the city.*
>
> *If I am a knave, then the right fork will take you to the city.*

What should you do?

Solution: Let us consider in turn the three possibilities: Fred is fully knave, Fred is a neutral, and Fred is fully knight.

Fred cannot be fully knave, for then the antecedent of his first conditional would be false. This would make the entire conditional true. This is a contradiction, since people who are fully knave cannot make true statements.

Fred also cannot be a neutral. If he were, then the antecedents of both of his conditional statements would be neutral. Table 13.3 indicates that conditional statements with neutral antecedents can be neutral only when their conclusions are false. However, since we know that one of the paths leads to the city, it cannot be that both conclusions are false. We have reached another contradiction.

It follows that Fred is fully knight. Since his first conditional is now seen to have a true antecedent, its conclusion must be true as well. So you should take the left fork. (His second conditional has a false antecedent, which implies the whole statement is true regardless of the falseness of the conclusion.) □

Here is a puzzle to illustrate the properties of three-valued conjunctions and disjunctions. It also illustrates that three-valued logic forces us to reconsider our ideas about contradictions and tautologies.

Puzzle 69. *You meet three people who make the following statements:*

> Gina : *Iris is a knight and Iris is a knave.*
>
> Hank : *Iris is a knight or Iris is a knave.*
>
> Iris : *Gina is a knight or Hank is a knave or I am a neutral.*

What can you conclude about Gina, Hank, and Iris?

Solution: Gina's statement is a classical contradiction, and Hank's statement is a classical tautology. On a classical island, we would know immediately that Gina is a knave and Hank is a knight. On a three-valued island, it is possible that either or both statements could be neutral. However, even on a three-valued island, there is no scenario in which

Gina's statement is true or Hank's statement is false. This can be seen by systematically checking all the cases.

We conclude that Gina is either neutral or fully knave, while Hank is either neutral or fully knight. It follows that neither of the first two parts of Iris's statement are true. (Both parts are either false or neutral.) Let us now consider the possibilities for Iris's type.

If Iris is fully knight, then her disjunction must be true. Since neither of the first two parts are true, we must have that the third part is true. But the third part is actually false in this scenario, so this option does not work.

If Iris is a neutral, then her disjunction must be neutral. But in this scenario, the third part of her disjunction is true. That makes the whole statement true, so this option does not work either.

We conclude that Iris is fully knave. This immediately implies that Hank's statement is true and Gina's statement is false. So Hank is fully knight and Gina is fully knave. ☐

If you feel like you have the hang of this, have a go at the puzzles in Section 13.5.

13.3   The Fuzzy Island

The second of the two islands mentioned in Section 13.1 took a different approach. The local logicians confronted the same problem of vagueness as the three-valued island, and they agreed that the principle of bivalence was too constraining. However, they opted for a different solution.

In their view, truth was best viewed as a matter of degree. The idea was that the statement "He is a knight" is more true when directed at a knight who had just entered his transitional phase than when directed at a knave who had just entered his. For that reason, they employed a continuum of truth values. A truth value, for them, was a real number between 0 and 1 inclusive. A truth value of 0 indicated complete falsity, while 1 indicated complete truth. If the statement "Joe is a knight" is assigned a truth value of, say, 0.9, then that would indicate that Joe was recently a knight but is now in the early stages of his transition. (So that he is mostly knight and just a little bit knave.) Since this approach suggested that truth was sometimes an imprecise concept, we shall refer to such islands as "fuzzy islands."

As on the three-valued island, not just any old statement can be assigned a truth value other than 0 or 1. Again, for nonvague statements, truth and falsity continue to mean what they have always meant. Intermediate truth values are used only in the case of ambiguous statements.

In particular, they apply only to assignments of knighthood or knavehood to those in the transitional state (or to compound statements built from those atoms through the standard connectives).

If we let P and Q denote arbitrary propositions, and if we denote by $v(P)$ and $v(Q)$ their truth values, then the islanders employ the following conventions for assigning truth values to compound statements:

$$v(\neg P) = 1 - v(P),$$

$$v(P \wedge Q) = \min(v(P), v(Q)),$$

$$v(P \vee Q) = \max(v(P), v(Q)).$$

When you see the notation $v(P)$, you should mentally say, "The truth value of P," or more simply, "v of P."

In words, the truth values assigned to a statement and its negation must sum to 1. A conjunction is as true as its least-true conjunct, while a disjunction is as true as its most-true disjunct. All three of these conventions agree with the dictates of classical logic when only classical truth values (1 or 0) are employed. As it happens, they also agree with three-valued logic if we use 1, 1/2, and 0 for true, neutral, and false, respectively.

The convention for conditionals is more complex:

$$v(P \rightarrow Q) = \begin{cases} 1 & \text{if } v(P) \leq v(Q), \\ 1 - (v(P) - v(Q)) & \text{if } v(P) > v(Q). \end{cases}$$

The idea is that if P is less true than Q, then we declare the conditional P → Q to be true (have truth value 1). This is in agreement with the rule in classical logic. If P is more true than Q, then we view the conditional as defective in some sense. The extent to which it differs from perfect truth is found by subtracting from 1 the magnitude of the drop in truth value from P to Q. Thus a conditional statement whose antecedent is far more true than its conclusion should be regarded as mostly false.

The transitional phase also complicates the island's sociology. A recent knight who was less than halfway through his transitional phase is referred to as a "quasiknight." At this stage, he only makes statements with high truth values. Near the end of his transition, he becomes a "quasiknave." In this condition he only makes statements with low truth values. Specifically, if P is a proposition spoken by a person A, then we have:

$$\begin{cases} v(P) = 1 & \text{if } A \text{ is a knight,} \\ 0.5 < v(P) < 1 & \text{if } A \text{ is a quasiknight,} \\ 0 < v(P) < 0.5 & \text{if } A \text{ is a quasiknave,} \\ v(P) = 0 & \text{if } A \text{ is a knave.} \end{cases}$$

Moreover, if P is the proposition "*A* is a knight," then the above four cases again provide the possible values for $v(P)$. Note that no islander ever makes statements whose truth value is exactly 0.5. (This is a convention adopted because it leads to cleaner puzzles. It is not that a statement cannot in principle have a truth value of 0.5; it is simply that on this particular island, no one ever utters such statements.)

Let us make explicit a few principles that will be useful for the puzzles to come.

- Earlier I said that on fuzzy islands, knights only make true statements and knaves only make false statements. That should be understood to mean that knights only make statements with truth value 1, while knaves only make statements with truth value 0. In referring to such people, let us continue to use the expressions "fully knight" and "fully knave."
- The propositions "Joe is not a knight" and "Joe is a knave" should be understood as logically equivalent. They always have the same truth value. For that reason, we can say that if P is the proposition "Joe is a knight," then ¬P is the proposition "Joe is a knave." The same applies to the pair of statements "Joe is not a knave" and "Joe is a knight."
- In reasoning through the puzzles to come, it will help to think of a quasiknight as someone who is more knight than knave. Likewise, you should think of a quasiknave as someone who is more knave than knight.
- There is no vagueness regarding one's status as a quasiknight or a quasiknave. That is, the propositions "Joe is a quasiknight" and "Joe is a quasiknave" can only have truth values 0 or 1. If Joe was recently a knight but has just entered his transitional phase, then the statement "Joe is a quasiknight" has truth value 1, but the statement "Joe is a knight" is vague and has a truth value somewhere strictly between 0.5 and 1. The same is true if the word "knight" is replaced with "knave," except that now the statement "Joe is a knight" has a truth value strictly between 0 and 0.5.

It is only the heartiest of travelers who would converse with the islanders in the hope of gaining information. To appreciate the difficulties, let us try some puzzles.

Puzzle 70. *What can you conclude from the following dialogue?*

> Joan : *Kent is a quasiknight.*
>
> Kent : *Joan is a quasiknave.*

Solution: Both Joan and Kent have made statements that can only have truth values 0 or 1. It follows that each is either fully knight or fully knave. Therefore, Kent is not a quasiknight and Joan is not a quasiknave. We conclude that Joan and Kent have both made statements with truth value 0, which implies that both are fully knave. □

Puzzle 71. *What can you conclude from the following dialogue?*

> *Lynn* : *Myra is a knight.*
>
> *Myra* : *Lynn's statement has truth value 0.8.*

Solution: Myra's statement is of a sort that can only have truth value 1 or 0. This is because Lynn's statement has a definite truth value, and that value is either 0.8 or it is not. If we suppose that Myra's statement has truth value 1, then she must be fully knight. It would then follow that Lynn's statement has truth value 1, which would make her fully knight as well. Since people who are fully knight only make statements with truth value 1, this would imply that Myra's statement actually has truth value 0, which is a contradiction.

It follows that Myra's statement has truth value 0, and therefore that she is fully knave. This implies that Lynn's statement also has truth value 0, meaning that she is fully knave as well. So Lynn and Myra are both fully knave. □

Here are two puzzles involving conjunctions and disjunctions.

Puzzle 72. *Suppose I tell you that my friend Nick, a resident of the fuzzy island, said "I am a knave or a quasiknight." What would you conclude?*

Solution: Let P be the proposition "Nick is a knave," and let Q be the proposition "Nick is a quasiknight." Then Nick's statement is equivalent to P ∨ Q.

Now, $v(Q) = 0$ or 1. If $v(Q) = 1$, then Nick really is a quasiknight. But then we would have

$$v(P \vee Q) = \max(v(P), v(Q)) = 1,$$

which implies that Nick is fully knight. This is a contradiction.

It follows that $v(Q) = 0$ and therefore that $v(P \vee Q) = v(P)$. If Nick is fully knight, then $v(P) = 0$. This is a contradiction, since it entails that someone who is fully knight has made a statement with truth value 0. We get a similar contradiction if we suppose Nick to be fully knave.

The trouble is that we also get a contradiction if we suppose Nick to be a quasiknave. For then we would have that $0.5 < v(P) < 1$ since a quasiknave is more knave than knight. But this would imply that $v(P \lor Q) > 0.5$, which is too high for a quasiknave.

So you would conclude that I had lied to you, since no islander can make the statement I attributed to Nick! □

Let us consider some puzzles based on conditional statements.

Puzzle 73. *One day you meet Oren, who says, "If I am a knight, then I am a knave." What can you conclude about Oren?*

Solution: Let P denote the statement "Oren is a knight." To simplify the notation, let us define Q to be ¬P. That is, Q is the statement "Oren is a knave," and Oren's statement is then equivalent to P → Q.

Now, if Oren is a knave or a quasiknave, then $v(P) < v(Q)$, so that $v(P \to Q) = 1$, which is too high. If Oren is fully knight, then $v(P) > v(Q)$, which implies that $v(P \to Q) < 1$, which is too low.

It follows that Oren is a quasiknight, but we can also conclude a bit more. Since $v(P) > v(Q)$, we have that

$$v(P \to Q) = 1 - (v(P) - v(Q)).$$

Moreover, we must also have $0.5 < v(P \to Q) < 1$. This is only possible if $v(P) - v(Q) < 0.5$, which in turn implies that $0.5 < v(P) < 0.75$. Thus, Oren is a quasiknight who is between a quarter and half way through his transition. □

In the next puzzle, it is convenient to refer to the "type" of an islander, that is, fully knight, fully knave, quasiknight, or quasiknave.

Puzzle 74. *What can you conclude from the following dialogue?*

Pete: Quin is a knight. Also, Rory is a knave.

Quin: If I am a knave, then Rory is a quasiknight.

Rory: If I am a quasiknave, then Pete and Quin are of the same type.

Solution: The most efficient solution to this problem involves the following observation. Suppose that P and Q are propositions with $v(Q) = 0$. It then follows that $v(P) \geq v(Q)$. We can now carry out the following calculation:

$$v(P \to Q) = 1 - (v(P) - v(Q)) = 1 - v(P) = v(\neg P).$$

In the present case, define P to be the proposition "Quin is a knave," and Q to be the proposition "Rory is a quasiknight." Then Pete's first statement is ¬P, while Quin's statement is equivalent to P → Q. Let us also define R to be the proposition "Rory is a quasiknave."

Notice that if $v(R) = 1$, then $v(Q) = 0$. Also, it follows from our observation that if $v(Q) = 0$, then $v(P \to Q) = v(\neg P)$, which implies that Pete and Quin are of the same type. We conclude that if Rory is a quasiknave, then Pete and Quin are the same type, which is precisely what Rory said. It follows that Rory's statement has truth value 1, and therefore that Rory is fully knight.

This shows that Pete's second statement has truth value 0, implying that Pete is fully knave. Since Pete and Quin are of the same type, we conclude that Quin is fully knave as well. So the solution is that Pete and Quin are fully knave, while Rory is fully knight. □

The foregoing puzzles supply a solid foundation for understanding how fuzzy conditionals work. Have a go at the additional puzzles provided in Section 13.5.

13.4 Modus Ponens and Sorites

We saw in Puzzle 69 that familiar notions like tautologies and contradictions must be reconsidered when employing nonclassical logic. Also in need of reconsideration are familiar ideas about validity. Even something as fundamental as *modus ponens* (by which is meant the argument form in which Q is said to follow from assuming the truth of P and P → Q) must be reevaluated. How should we understand validity in the context of fuzzy logic?

In classical logic, we say an argument is valid if the conclusion must be true when all of the premises are assumed to be true. In fuzzy logic, "truth" is a matter of degree. Informally, then, we might say that an argument in this context is valid if the conclusion must have a high truth value given that all of the premises have high truth values. This can be made more precise by deciding on a set of "distinguished" truth values and then declaring an argument to be valid if the conclusion must have a distinguished truth value when all of the premises have distinguished truth values. For example, we might decide, arbitrarily, that a proposition P has a distinguished truth value if $0.8 \leq v(P) \leq 1$.

This choice seems reasonable, but our next puzzle shows that it has some surprising consequences.

Puzzle 75. *You are observing a group of school kids. One of them, Sara, approaches you and makes the following statements about some of her classmates:*

1. *Todd is a knight.*
2. *If Todd is a knight, then so is Ursa.*
3. *If Ursa is a knight, then so is Vera.*

At this point, Vera runs up to you and says, "Don't believe Sara! She's a knave!" What can you conclude?

Solution: On a classical island, this conversation would be impossible. For suppose Sara is fully knight. In this case, all three of her statements have truth value 1. A straightforward application of *modus ponens* would now show that Vera must also be fully knight. In this scenario, however, Vera's statement has truth value 0. This is a contradiction.

Now suppose Sara is fully knave. In this case, her first statement has truth value 0. But then her second statement is a conditional with a false antecedent, which implies the whole statement is true. This, again, is a contradiction.

This would exhaust the possibilities on a classical island. On fuzzy islands, however, we have two additional possibilities. The first is that Sara is a quasiknave. That this is impossible is shown by the following argument: If Sara is a quasiknave, then her first statement has a truth value that is no greater than 0.5. But then her second statement is a conditional whose antecedent has a truth value no greater than 0.5. Now, the conditional statement $P \to Q$ has truth value 1, when $v(P) \leq v(Q)$, or truth value $1 - (v(P) - v(Q))$, when $v(P) > v(Q)$. So if $v(P) < 0.5$, then $v(P \to Q) > 0.5$. This is impossible if Sara is a quasiknave.

Can Sara be a quasiknight? This might seem impossible at first. Reasoning informally, we might say that since Sara is a quasiknight, all of her statements have high truth values. An application of *modus ponens* would then suggest that the statement "Vera is a knight" must also have a high truth value. But this is not the case, since Vera's statement has truth value 0 in this scenario.

This argument is a bit *too* informal, however. It is possible for Sara to be a quasiknight, but only if we have something like the following scenario. We start with some definitions:

P is the proposition "Todd is a knight."

Q is the proposition "Ursa is a knight."

R is the proposition "Vera is a knight."

Sara's statements are then, sequentially, P, $P \to Q$, and $Q \to R$. Now suppose we have something like $v(P) = 0.7$, $v(Q) = 0.4$, and $v(R) = 0.1$. We would then compute:

$$v(\text{P} \to \text{Q}) = 1 - (0.7 - 0.4) = 0.7$$
$$v(\text{Q} \to \text{R}) = 1 - (0.4 - 0.1) = 0.7.$$

These truth values are in the proper range for a quasiknight.

We conclude that Sara is a quasiknight. It now follows immediately from her first assertion that Todd is also a quasiknight. Vera, for her part, must be a quasiknave, since her assertion that Sara is a knave is now seen to be mostly, but not completely, false.

Ursa's status, however, is more ambiguous. In the scenario I just outlined, our assumption regarding the truth of Q implies that Ursa is a quasiknave. However, it would also have been consistent to assume that $v(\text{P}) = 0.7$, $v(\text{Q}) = 0.51$, and $v(\text{R}) = 0.3$. In this case, if you care to check the calculations, Ursa would be a quasiknight.

Thus, the solution to the problem is that Sara and Todd are quasiknights, Vera is a quasiknave, and Ursa is a quasi-something. □

This puzzle shows, surprisingly, that *modus ponens* is no longer a valid argument form in fuzzy logic. More precisely, we have shown that each link in a chain of implications can have a high truth value, while the conclusion of the final item has a low truth value.

Perhaps, though, this possibility should be viewed as a feature, and not a bug, of fuzzy logic. Classical logic, you see, faces the problem of "sorites" paradoxes. (Note that this use of the term "sorites" is different from our use of it in Chapter 5.) The problem occurs when extreme ends of a continuum differ in some respect, but it is not possible to find a clear point on the continuum where the difference first manifests itself. The classic example involves the notion of a heap ("sorites" derives from the Greek word for "heap"). One grain of sand does not make a heap. Surely, though, adding one grain of sand to something that is not a heap cannot suddenly make it a heap. We conclude that two grains of sand is also not a heap. But now iterating the argument leads to the conclusion that no number of grains of sand can make a heap, which is plainly absurd.

The next puzzle shows how positing a continuum of truth values might evade the sorites problem. Assume that an islander's transition from fully knight to fully knave happens continuously, but very gradually. More precisely, let us make two assumptions. The first is that the full transition occurs at a constant rate over a period of several days. The second is that for any time interval t, no matter how small, a person in the transitional phase is closer to the end of her transition after t has elapsed than she was before t has elapsed.

Puzzle 76. *You meet Wren, who tells you about her friend Xena. Wren makes the following statements:*

1. *Xena is a knight.*
2. *If anyone is currently a knight, then she will still be a knight 1 second from now.*
3. *Therefore, Xena will still be a knight 1 second from now.*

What can you conclude about Wren and Xena? Is Wren's argument valid?

Solution: A knight who recently entered the transitional phase will be very slightly less of a knight 1 second from now than she currently is. It follows that Wren's conditional statement has a conclusion whose truth value is very slightly smaller than the truth value of the antecedent. Thus, the whole conditional has a truth value that is very slightly less than 1, which implies that Wren is a quasiknight. This now implies that Wren's first statement likewise has a truth value between 0.5 and 1, which implies that Xena is a quasiknight as well.

The argument is valid in the sense that when the premises have high truth values, the conclusion does as well. It might seem, therefore, that this is an instance of a sorites paradox, since iterating the argument leads to the conclusion that Xena will always be a knight, when we know that, in reality, after some number of seconds, she will be fully knave. In this case, however, it is straightforward to resolve the paradox. Each time we iterate the argument, the conclusion is slightly less true than it was the previous time. Therefore, after some specific number of iterations, the conclusion will, for the first time, cease to have a distinguished truth value. After this number of iterations, the argument will cease to be valid. □

The ability to defuse sorites paradoxes has historically been an important motivation for many-valued and infinite-valued logics. By recognizing a continuum of truth values, we can claim that each link in the chain of implications in a sorites argument is less true than the one before, which implies that the chain is broken at some specific point.

13.5 Puzzles for Solving

Now it is your turn to have a go! See what you can make of the following puzzles. The first three take place on the three-valued island.

We start with two warm-ups, to see whether you have the basics.

TABLE 13.4.
The truth table for biconditional
statements in three-valued logic.

$\rightarrow$	T	N	F
T	T	N	F
N	N	T	N
F	F	N	T

Puzzle 77. *What can you conclude from the following dialogue?*

> *Yoav*: *Neither of us is a neutral.*
>
> *Zach*: *Actually, Yoav is a neutral.*

Puzzle 78. *Explain why the following dialogue is impossible:*

> *Alan*: *Barb is a knight or a neutral.*
>
> *Barb*: *Alan is a knave or a neutral.*

Now try something a bit more complex:

Puzzle 79. *What can you conclude from the following dialogue?*

> *Cora*: *Doug is a neutral, and Ezra is a knave.*
>
> *Doug*: *If Ezra is a knight, then Cora is a neutral.*
>
> *Ezra*: *Cora is a knight.*

Our next puzzle includes two people who make "if and only if" statements, which are sometimes referred to as biconditionals. Recall that to define a biconditional, we employ the classical identity

$$P \leftrightarrow Q \equiv (P \rightarrow Q) \wedge (Q \rightarrow P).$$

The rules for three-valued connectives defined previously can be used to construct the proper table, shown in Table 13.4.

If you have absorbed that, and if you really want to go for glory, then you can take a shot at the final three-valued puzzle.

Puzzle 80. *One day a visitor encountered eight people. Some of them were normals who were visiting from another island. Normals sometimes made true statements and sometimes made false statements, but they never made neutral statements. The following dialog ensued:*

> Faye : If Mark is not a neutral, then neither is Lara.
>
> Gail : Lara is a knight, and Kyle is a knave.
>
> Herb : Jack is a knight if and only if Ivan is a neutral.
>
> Ivan : Gail is a neutral if and only if Faye is not a normal.
>
> Jack : If Kyle is a knight, then Lara and Herb are normals.
>
> Kyle : Faye is a neutral, or Mark is a neutral.
>
> Lara : Gail is a knave, and Herb is a knave.
>
> Mark : I am not a neutral.

Well, the poor visitor could not make heads or tails out of all that. But then someone he knew to be fully knight told him that among the eight people were exactly two knights, two knaves, two normals, and two neutrals. "Eureka!" cried the visitor. "Now I know exactly who is who!"

Can you do as well as the visitor? Determine the types of all eight people.

We now move on to a few puzzles that take place on the fuzzy island. As before, we begin with a warm-up to cement the basics.

Puzzle 81. *While walking on the fuzzy island, you meet Neil. He says, "There is no type of person who can assert that I am a quasiknave." What would you conclude about Neil?*

In the next puzzle, it will be convenient to refer to the "rank" of an islander. The idea is that the closer someone is to being fully knight, the higher their rank. Someone who is fully knight is of higher rank than a quasiknight, who in turn is of higher rank than a quasiknave. Someone who is fully knave is of the lowest rank.

Puzzle 82. *What can you conclude from the following dialogue?*

> Olaf : Quan is a knave.
>
> Paul : Quan is a knave, and Olaf is a knight.
>
> Quan : Olaf is of lower rank than Paul.

Let us close with a puzzle that requires a more detailed analysis than what has come before.

Puzzle 83. *What can you conclude from the following dialogue?*

> Rene : *I am a knight, and Stan is a knave. Tara is a knave.*
>
> Stan : *If Rene is a knight, then I am a knave.*
>
> Tara : *Stan is a quasiknight, or Rene is a knave.*

13.6   Solutions

Puzzle 77. Both Yoav and Zach have made statements that have classical truth values. So neither is a neutral. This immediately implies that Yoav's statement is true and Zach's statement is false. Therefore, the solution is that Yoav is fully knight and Zach is fully knave.

Puzzle 78. To see that this dialogue is impossible, we consider in turn Alan's possible types. If Alan is fully knight, then Barb's statement is false. This would imply that she is a knave. But in this case, Alan's statement would be false, which is a contradiction. Now suppose Alan is a neutral. In this case, Barb's statement is seen to be true, which would imply she is fully knight. But in this case, Alan's statement would be true, and not neutral, which is another contradiction. Finally, suppose Alan is a knave. Then Barb's statement is true, which implies that she is fully knight. But this would make Alan's statement true, which would be another contradiction. Since we have run out of possibilities, we conclude that the dialogue is impossible.

Puzzle 79. The challenge of solving a problem of this sort is finding a way in. One could make some sort of arbitrary starting assumption and hope that it leads somewhere helpful. This can be quite challenging to pull off, though, since on a three-valued island, there are always many possible cases to consider. So we have to be a bit savvy about finding a good starting point.

A helpful approach is to focus on statements that cannot be neutral. In this case, the second part of Doug's conditional statement attracts attention, since it must have a classical truth value. Recall that a conditional statement whose consequent has a classical truth value can only be neutral when the antecedent is neutral and the consequent is false. That means that if Doug is a neutral, then we immediately have that Ezra is a neutral and Cora is not a neutral. But if Cora is not a neutral, then Ezra's statement has a classical truth value. We have reached a contradiction, so we conclude that Doug is not a neutral.

Everything else unravels quickly. The first part of Cora's conjunction is now seen to be false. That implies all by itself that her entire statement is false, and it follows that Cora is a knave. Now we know Ezra's statement is false, which implies he is a knave as well. So Doug's statement is seen to be true, and, therefore, he is a knight.

Thus the solution is that Cora is a knave, Doug is a knight, and Ezra is a knave.

Puzzle 80. With such a wealth of statements to consider, we need to look carefully for a way in. As before, focus on statements that, by their natures, cannot be neutral.

Consider Mark, for example. He cannot be a neutral, since his statement clearly has a classical truth value. He also cannot be a knave, since in that case, his statement would be true. It follows that he is a knight or a normal.

Now move on to Faye. She is also not a neutral, since both parts of her conditional statement have classical truth values. Under these circumstances, a three-valued conditional statement cannot be neutral, as shown by Table 13.3.

What about Kyle? Both parts of his disjunction have classical truth values, implying that the whole statement does as well. Moreover, since we know that neither Faye nor Mark is a neutral, we see that Kyle's statement is actually false. This shows that Kyle is either a knave or a normal.

We have already made considerable progress! Our findings to this point are summarized in Table 13.5.

Turn now to Jack. We have already established that Kyle is neither a knight nor a neutral. The antecedent of Jack's conditional is therefore false, which implies that his whole statement is true. Thus he is either a knight or a normal.

TABLE 13.5.
First steps toward a solution of
Puzzle 80.

Person	Possible Type
Faye	knight, knave, normal
Kyle	knave, normal
Mark	knight, normal

TABLE 13.6.
Further progress toward a solution of
Puzzle 80.

Person	Possible type
Herb	knight, knave, normal
Ivan	knight, knave, normal
Jack	knight, normal

Ivan is also not a neutral, because both parts of his biconditional have classical truth values.

And Herb cannot be a neutral. The second part of his biconditional statement has a classical truth value. Moreover, we also know that Jack is either a knight or a normal. So the first part of Herb's statement has a classical truth value. Under these circumstances, his whole statement must have a classical truth value.

These new results are summarized in Table 13.6. That makes six people who cannot be neutrals. We conclude that Lara and Gail must be the two neutrals.

Let us now turn our attention to what these neutrals said. Each of their statements is a conjunction whose first part is neutral. Notice that the second parts of their statements must have classical truth values, since neither Kyle nor Herb is a neutral. In each case, however, the entire statement must be neutral. This is possible only if the second part is true in both cases. It follows that Kyle and Herb are both knaves.

We have now discovered the types for half of the speakers: Kyle and Herb are the knaves, while Lara and Gail are the neutrals. There are two knights and two normals among the remaining four. Reconsidering Faye's statement, we now see that her conditional has a true antecedent and false consequent. Therefore, her statement is false. Since the knaves are accounted for, we conclude that Faye must be a normal.

Herb's statement must be false, since he has been established as a knave. The second part of his statement is also false, since Ivan is not a neutral. It follows that the first part must be true, which implies that Jack is a knight.

That leaves only Ivan and Mark. Each is either a knight or a normal. Look at Ivan's statement. Since Gail really is a neutral, while Faye is actually a normal, we can evaluate his statement as false. Since he cannot be a knave, we conclude that he is a normal. This forces Mark to be a knight, and we are done.

TABLE 13.7.
The full solution of Puzzle 80.

Type	People
knight	Jack, Mark
knave	Kyle, Herb
normal	Faye, Ivan
neutral	Lara, Gail

The full solution is shown in Table 13.7.

Puzzle 81. If Neil is actually a quasiknave, then a knight could identify him as such. If Neil is not a quasiknave, a knave could identify him as such. We must conclude that Neil's statement has truth value 0, which implies that he is a knave.

Puzzle 82. Let us define P to be the proposition "Quan is a knave" and define Q to be the proposition "Olaf is a knight." Since

$$v(P \land Q) = \min(v(P), v(Q)) \leq v(P),$$

we see that the truth value of Olaf's statement is not smaller than the truth value of Paul's statement. It follows that Olaf's rank is not lower than Paul's rank, which implies that Quan's statement has truth value 0. From this we quickly see that both Olaf's statement and Paul's statement have truth value 1.

So the solution is that Olaf and Paul are fully knight, while Quan is fully knave.

As an aside, note that if in Puzzle 82, we change Paul's statement to "Quan is a knave, or Olaf is a knight," and change Quan's statement to "Olaf is of higher rank than Paul," then the solution would be essentially unchanged. In this case, we would have

$$v(P \lor Q) = \max(v(P), v(Q)) \geq v(P),$$

which would imply that Olaf cannot possibly be of higher rank than Paul.

Puzzle 83. We have to be systematic when considering the various cases. Let us make the following definitions:

P is the proposition: "Rene is a knight."

Q is the proposition: "Stan is a knave."

Then Rene's first statement is equivalent to $P \land Q$. Stan's statement is equivalent to $P \rightarrow Q$.

We can now distinguish two cases.

1. Suppose $v(Q) \geq v(P)$. In this case, the consequent of Stan's conditional has a truth value that is not smaller than its antecedent. This implies that $v(P \to Q) = 1$. It follows that Stan is fully knight, which immediately implies that $v(Q) = 0$. Since we are assuming $v(Q) \geq v(P)$, we must also have that $v(P) = 0$, implying that Rene is fully knave.

2. Suppose $v(Q) < v(P)$. Given this assumption, we immediately conclude that $v(Q) < 1$. Since

$$v(P \wedge Q) = \min(v(P), v(Q)) = v(Q) < 1,$$

we know that Rene is not fully knight. We now distinguish three further subcases:

(a) Suppose that Rene is a quasiknight. The analysis now proceeds in a manner similar to the first case. Since Rene is assumed to be a quasiknight, we have that

$$0.5 < v(P \wedge Q) < 1.$$

Also, we are still assuming that $v(Q) < v(P)$. It then follows that

$$0.5 < \min(v(P), v(Q)) < 1,$$

which implies that

$$0.5 < v(Q) < v(P) < 1.$$

Consequently, $v(P) - v(Q) < 0.5$. Therefore,

$$0.5 < v(P \to Q) < 1.$$

This would again imply that Stan is a quasiknight. So Rene and Stan are both quasiknights in this scenario.

(b) Suppose that Rene is a quasiknave. Then

$$0 < v(P \wedge Q) < 0.5.$$

It follows that

$$0 < \min(v(P), v(Q)) < 0.5.$$

Since we are assuming $v(Q) < v(P)$, we conclude that $0 < v(Q) < 0.5$. But since Rene is assumed to be a quasiknave, we also have that $0 < v(P) < 0.5$. Consequently, since we know that

$$v(P \to Q) = 1 - (v(P) - v(Q)),$$

we have that $0.5 < v(P \to Q) < 1$. This implies that Stan is a quasiknight. Thus, in this scenario, Rene is a quasiknave and Stan is a quasiknight.

(c) Finally, suppose that Rene is fully knave. Then $v(P) \land Q = 0$. Also, $v(P) = 0$. Since the antecedent of Stan's conditional has truth value 0, we have that $v(P \to Q) = 1$. Thus Stan is fully knight.

Our analysis has revealed that there are two possible scenarios. One is that Rene is fully knave and Stan is fully knight. The other is that Stan is a quasiknight and Rene is a quasi-something. Thus either Stan is a quasiknight or Rene is fully knave.

But this is precisely what Tara said! It follows that her statement has truth value 1, and therefore that she is fully knight. Thus, Rene's second statement has truth value 0, implying that he is fully knave. This immediately implies that Stan is fully knight.

So the solution is that Rene is fully knave, while Stan and Tara are both fully knight.

PART V
Miscellaneous Topics

CHAPTER 14

The Saga of the Hardest Logic Puzzle Ever

The Hardest Logic Puzzle Ever (HLPE) was introduced by George Boolos, then a philosophy professor at the Massachusetts Institute of Technology, in a 1996 paper. The paper was titled "The Hardest Logic Puzzle Ever." Since then, the puzzle has attracted an impressive academic literature, which we shall survey in this chapter. When we are done, I am sure you will agree that the term "saga" is appropriate.

The HLPE is an especially fiendish version of a question puzzle, which we discussed in Chapter 11. Some of the techniques developed in that chapter are about to reappear.

14.1 Boolos Introduces the Puzzle

Here is George Boolos's statement of the puzzle:

> Three gods A, B, C are called, in some order, True, False, and Random. True always speaks truly, False always speaks falsely, but whether Random speaks truly or falsely is a completely *random* matter. Your task is to determine the identities of A, B, and C by asking three yes-no questions; each question must be put to exactly one god. The gods understand English, but will answer all questions in their own language, in which the words for "yes" and "no" are "da" and "ja," in some order. *You do not know which word means which*. (Boolos 1996, 62)

Note that "True," "False," and "Random" should be taken to be the actual names of the gods. Thus, we do not need to refer to the god who speaks truthfully or the god who speaks falsely, but can instead just call them True, False, and Random. Boolos credits Raymond Smullyan with the general outline of the puzzle, and computer scientist John McCarthy for adding the part about not knowing the meanings of "da" and "ja." Smullyan later denied having invented it, though he did discuss puzzles of a similar nature in several of his books.

After stating the puzzle, Boolos (1996, 63) provides four bullet points to clarify his intent:

- It could be that some god gets asked more than one question (and hence that some god is not asked any question at all).
- What the second question is, and to which god it is put, may depend on the answer to the first question.
- Whether Random speaks truly or not should be thought of as depending on the flip of a coin hidden in his brain: if the coin comes down heads, he speaks truly; if tails, falsely.
- Random will answer da or ja when asked any yes-no question.

The third of these bullet points will have special relevance in Section 14.3.

Even in light of Boolos's clarifications, some ambiguity remains with regard to Random's behavior. Does he toss the coin anew every time he is asked a question? Or does he toss the coin once "per session," as it were (so that if he is asked two questions, he replies as True for both or False for both)? It is clear from Boolos's solution, and from the way later writers construed Random's behavior, that we should assume he flips the coin anew each time he is asked a question. Moreover, after each question, he immediately flips the coin in preparation for the next question. In this way, we can say that at any given moment, Random is either in the mental state True or in the mental state False.

Boolos's paper is four pages long. It was published in *The Harvard Review of Philosophy*, which strives to appeal to a general audience, as opposed to a regular research journal, which is directed at professionals. There is no indication that he intended to provide anything to readers other than an amusing recreation. Nonetheless, his brief paper has spawned a small industry of scholarly work, in which ever-more ingenious solutions are provided, and increasingly difficult versions of the puzzle are devised.

The remainder of this section is devoted to Boolos's solution. Since this solution is quite complex, let us follow his lead of building up to it through a sequence of three simpler puzzles.

Here is Boolos's first warm-up puzzle:

Puzzle 84. *Noting their locations, I place two aces and a jack face down on a table, in a row; you do not see which card is placed where. Your problem is to point to one of the three cards and then ask me a single yes-no question, from the answer to which you can, with certainty, identify* *one of the three cards as an ace. If you have pointed to one of the aces, I will answer your question truthfully. However, if you have pointed to the jack, I will answer your question yes or no, completely at random. (Boolos 1996, 63)*

Solution: You can point to the middle card and ask, "Is the left card an ace?"

Suppose you are pointing to one of the aces. Then I will answer your question truthfully. If I say yes, then the left card *is* an ace, and if I say no, then the right card must be the other ace.

Now suppose you are pointing to the jack. Then I will answer randomly. However, in this case, the left and right cards will both be aces. So it will again be true, if only in a trivial sense, that a "yes" answer implies the left card is an ace and a "no" answer implies the right card is an ace.

Either way, we will accomplish our goal by interpreting a "yes" answer to mean "Choose the left card," and a "no" answer to mean "Choose the right card." □

Clever! Sadly, things are about get considerably more complex. Let us move on to the second puzzle.

Puzzle 85. *Suppose that, somehow, you have learned that you are speaking not to Random, but to True or False—you don't know which—and that whichever god you're talking to has condescended to answer you in English. For some reason, you need to know whether Dushanbe is in Kirghizia. What one yes-no question can you ask the god from the answer to which you can determine whether or not Dushanbe is in Kirghizia? (Boolos 1996, 63)*

Kirghizia, if you are wondering, is the former name of the small country today known as Kyrgyzstan. Dushanbe is the capitol city of the country of Tajikistan. Therefore, Dushanbe is not in Kirghizia, though that fact is not relevant to what follows. That Boolos chose so obscure a question is suggestive of his sense of humor.

Now, if you have read Chapter 11, then this puzzle poses no challenge at all. It is just a straightforward application of the Nelson Goodman principle. Still, let us work through the solution formally.

Solution: To simplify the language, denote by X the assertion that Dushanbe is in Kirghizia. We now consider the question "Are you True if and only if X?" From our analysis in Section 11.4, we know that a "yes" answer implies that X is true, while a "no" answer implies that X is false. (Of course, X could represent any proposition at all without affecting the analysis.) □

Here is Boolos's third puzzle:

Puzzle 86. *You are now quite definitely talking to True, but he refuses to answer you in English and will only say "da" or "ja." What one yes-no question can you ask True to determine whether Dushanbe is in Kirghizia? (Boolos 1996, 63)*

This puzzle moves us into the territory covered by the generalized Nelson Goodman principle. Though we have analyzed situations like this before, it will be helpful to review how the cases play out.

Solution: Let us continue to use X to represent any assertion whose truth or falsity we seek to determine. In this case, the correct question is: "Does da mean yes if and only if X?" There are four cases to consider:

1. X is true and da means yes. Both parts of the biconditional are true, so the correct answer is yes. True will answer "da."
2. X is true and da means no. The correct answer is no, so True will answer "da."
3. X is false and da means yes. The correct answer is again no, so True will answer, "ja."
4. X is false and da means no. This time both parts of the biconditional are false, meaning the correct answer is yes. True will answer, "ja."

Our finding is that an answer of "da" implies that X is true, while an answer of "ja" implies that X is false. We have recorded these findings in Table 14.1. □

Boolos now combined these principles to solve the HLPE. His solution is most easily understood by working backward.

Suppose that we have somehow learned B's identity and that we know he is not Random. We could then ask him the question: "Does da mean yes if and only if A is Random?" There are two possible cases:

- Suppose we know that B is True. In this case, we can apply what we learned from Puzzle 86 to conclude that the answer "da" implies that A really is Random, while the answer "ja" implies that C is Random. Either way, we will definitively identify Random, from which the identity of the third god immediately follows.
- Suppose instead we know that B is False. If he answers "da," then we know True would have said "ja." This implies that A is not Random and therefore that C is. If instead he answers "ja," then we know True would have said "da." This would imply that A is Random. Again, either way, we will definitively identify Random, and therefore we can identify the remaining god as well.

Table 14.1.
The four cases relevant to the solution of Puzzle 86. Notice that True answers "da" when X is true and answers "ja" when X is false.

X Is	Da Means	Da Means Yes ↔ X	True's Response
True	Yes	True	Da
True	No	False	Da
False	Yes	False	Ja
False	No	True	Ja

Now back up one step. Imagine that we have identified one god as definitely not Random. He is either True or False, but we do not yet know which. We seek a question that will tell us which of the two he is. In this case, a question that works is: "Does da mean yes if and only if X?" where X is any assertion we know to be true. (Boolos suggests, "Does da mean yes if and only if Rome is in Italy?" for this purpose.) Since X is known to be true, an answer of da implies we are speaking to True, while an answer of ja implies we are speaking to False.

Now we back up one more step. We know how to proceed upon finding a god who is definitely not Random, and we have one question left to use. Can we find a question that will identify one of the gods as not Random?

The key is to combine the principles learned from the warm-up puzzles. If we ask either True or False the question: "Are you True if and only if B is Random?" then the answer will determine whether B is Random. We also know that the question, "Does da mean yes if and only if X?" will provoke True to answer "da" if X is true and "ja" if X is false, and we will get the reverse of this if the question is addressed to False.

What if we combine these ideas? Boolos suggests directing the following question to A: "Does da mean yes if and only if you are True if and only if B is Random?" That is very hard to parse, so let us do this: Define X to be the assertion: "You are True if and only if B is Random." Boolos's question now becomes: "Does da mean yes if and only if X?"

There are many cases to consider. Let us start by assuming that A is not Random. This narrows things down to eight possibilities:

- A is either True or False,
- B is either Random or not Random,
- Da means either yes or no.

Table 14.2 shows A's response in each scenario. The results are stark. If we know that A is not Random, then an answer of "da" implies that

TABLE 14.2.
The possible responses to Boolos's question under the assumption that A is not Random.

A's Identity	B's Identity	Da Means	A's Response
True	Random	Yes	Da
True	Random	No	Da
False	Random	Yes	Da
False	Random	No	Da
True	Not Random	Yes	Ja
True	Not Random	No	Ja
False	Not Random	Yes	Ja
False	Not Random	No	Ja

B is Random. An answer of "ja" implies that B is not random, which in turn implies that C is Random.

How did I fill in Table 14.2? If I had to analyze each of the eight cases separately, then things would quickly become tedious. Our basic principles, however, allowed me to fill in all eight rows by considering just two general cases.

Keep in mind that the question under discussion is: "Does da mean yes if and only if X?" Here X is: "You are True if and only if B is Random."

- Suppose A is True. From Puzzle 86, we know that a reply of "da" implies that X is true, while a reply of "ja" implies that X is false. Since A *is* True, we see that X is true when B is Random and false otherwise. Thus, if A is True, then "da" implies that B is Random, and "ja" implies that B is not Random.
- Suppose A is False. We then reverse our findings from Puzzle 86, so that "ja" implies that X is true, while "da" implies that X is false. Since A *is* False, we see that X is true when B is not Random, and X is false when B is Random. Thus, if A is False, then "da" implies that B is Random, and "ja" implies that B is not Random.

We are making progress, but what happens if A is Random? In this case, we must look to what we learned from Puzzle 84. If A is Random, *then neither B nor C is Random*!

It follows that if A now replies with "da," then we will not go wrong by assuming that C is not Random. Likewise, if A now replies with "ja," we will not go wrong by assuming that B is not Random.

In other words, regardless of whether A is True, False, or Random, we can interpret a reply of "da" to mean that C is not Random, and a reply of "ja" to mean that B is not Random. Either way, we definitively identify one of the gods as not being Random.

Putting everything together, here is Boolos's solution:

- First, ask A: "Does da mean yes if and only if (you are True if and only if B is Random)?" The answer to this allows you to conclude that one of B and C is not Random. For simplicity, assume that you have learned that B is not Random.
- Second, ask B: "Does da mean yes if and only if X?", where X represents any assertion you know to be true. Based on the reply, you will know for certain whether B is True or B is False.
- Third, ask B: "Does da mean yes if and only if A is Random?" The reply to this will allow you to identify A, from which C's identity will follow immediately.

Boolos closes his discussion by writing, "Well, I wasn't speaking falsely or at random when I said that the puzzle was hard, was I?" (Boolos 1996, 65). One wonders, though, if perhaps there is a simpler solution than the one Boolos devised.

14.2 Is There a Simpler Solution?

Boolos's choice for the second and third questions were essentially instances of the Nelson Goodman principle, and they hewed very closely to the reasoning employed by Smullyan. Had Boolos been aware of the very general version of the principle discussed in Section 11.5, he would have realized that the second and third questions pose no problem at all. The general principle allows us to get direct answers to our questions.

(I should mention that Boolos published his paper in 1996, while Smullyan did not publish his generalized Nelson Goodman principle until 2009. Smullyan made no mention of the HLPE in his 2009 book, which is interesting, since he was definitely aware of the puzzle.)

Specifically, following the terminology of Section 11.5, we can define a Type 1 native to be one who answers "da" when the correct answer is yes and answers "ja" when the correct answer is no. A Type 2 native is one who gives the reverse answers. Then, if we are addressing a god known not to be Random, we can ask: "Are you of Type 1 if and only if you are True?" For the reasons explained in Section 11.5, a response of "da" now indicates that we are speaking to True, while an answer of "ja" indicates we are speaking to False. One further direct question will then resolve the identities of the remaining gods.

It is Boolos's first question that introduces the real complexity into the proceedings. A nested biconditional question is not the easiest thing to work with, and we might seriously wonder whether some simpler question would suffice. Moreover, Boolos was not mindful of the distinction between a god who truly answers randomly, as opposed to a god who merely decides randomly whether to behave in the manner of the True god or the False god. This point will play a central role in Section 14.3.

In the meantime, however, let us consider whether biconditional questions are really necessary. It would be nice to get by with simpler questions, if that is possible. Such a solution was offered by Tim Roberts (2001). His solution to the puzzle was far simpler than the one given by Boolos. Whereas Boolos made use of complex, hard-to-parse biconditionals, Roberts employed only standard conditionals.

Recall that a conditional statement is false if its antecedent is true and its consequent is false. It is true in all other scenarios.

Roberts's approach employs the same general strategy as Boolos's solution. He uses the first question to definitively identify one of the gods as not being Random. The second question is then directed at this god, and serves to determine which he is among True and False. The final question is again directed at this god, and serves to resolve the identities of the other two gods.

Boolos built up to his solution by presenting three simpler puzzles of his own devising. Roberts dives right in, but he appears to take his inspiration from the Heaven/Hell puzzle of Section 11.3.

Roberts's solution begins by addressing god B and asking: "If I asked you if god A was Random, would you say 'da'?" The various cases can be divided into three groupings, depending on whether B is True, False, or Random. Let us consider these possibilities in turn.

First, suppose that B is True. There are four possibilities to consider, depending on whether A is False or Random, and depending on whether da means yes or no.

- Suppose that A is Random and da means yes. In this case, if True is asked whether A is Random, then he will reply "yes." Since da means yes, True would indeed reply "da" when asked whether A is Random. So B's answer to the conditional question is da.
- Suppose that A is Random and da means no. Again, True will answer "yes" when asked if A is Random. That means saying "ja." So B's answer to the conditional question is no (since he will not say "da" when asked if A is Random). Thus, B will again say "da" in this scenario.
- Now suppose that A is False and da means yes. True will answer "no" when asked if A is Random. That means saying "ja." So

TABLE 14.3.
The possible responses to Roberts's first
question, under the assumption that B is True.

A's Identity	Da Means	B's Response
Random	Yes	Da
Random	No	Da
False	Yes	Ja
False	No	Ja

TABLE 14.4.
The possible responses to Roberts's first
question, under the assumption that B is False.

A's Identity	Da Means	B's Response
Random	Yes	Da
Random	No	Da
True	Yes	Ja
True	No	Ja

B's answer to the conditional question will be no, and he will
answer "ja."

- Finally, suppose A is False and da means no. Again, True will answer
  "no" when asked if A is Random. That means saying "da." So B's
  answer to the conditional question is yes. That means saying "ja."

A summary of these four possibilities is recorded in Table 14.3.

Again, the results are stark. If B is True, then a response of "da" tells us
that A is Random, and a response of "ja" tells us that A is not Random.
Of course, if A is not Random, then C is.

Moving on, the possibilities when B is False are recorded in Table 14.4.
The analysis underlying this table is nearly identical to the case where B
is True. I explain lines one and four of the table, leaving the remaining
two lines as an exercise.

- Suppose A is Random and "da" means yes. When asked if A is Random, B will lie and say "no." That means replying "ja." So the correct
  answer to the conditional question is no. Answering correctly would
  then mean saying "ja." So B will lie and answer "da."

TABLE 14.5.
The possible responses to Roberts's second question, under the assumption that C is not Random.

C's Identity	Da Means	C's Response
True	Yes	Da
True	No	Da
False	Yes	Ja
False	No	Ja

- Suppose A is True and "da" means no. This time, B will say "yes" when asked if A is Random. That means replying "ja." So the correct answer to the conditional question is no. Answering correctly would then mean saying "da." So B will lie and answer "ja."

We conclude that if B is not Random, then a reply of "da" implies that A is Random and C is not Random. A reply of "ja" implies that A is not Random, and therefore C is Random. Moreover, if B is Random, then neither A nor C is Random. It follows that, in this case too, we will not go wrong in assuming that da implies that C is not Random, and that ja implies that A is not Random.

We have succeeded in our goal! With one question, we have identified a god who is definitely not Random. For simplicity, let us assume that this god is C.

The second question should then be addressed to C. We ask: "If I asked you if you always told the truth, would you say 'da'?" The possibilities are presented in Table 14.5. To avoid tedium, let us consider only the first and fourth lines of this table.

- Suppose that C is True and "da" means yes. When asked if he always tells the truth, C will reply "yes." That means saying "da." Thus, he will, indeed, say "da" when asked if he always tells the truth. So the correct answer to the conditional question is yes. C will give that answer by saying "da."
- Suppose that C is False and "da" means no. When asked if he always tells the truth, C will lie and say "yes." That means saying "ja." Thus, C will not reply "da" when asked if he always tell the truth. So the correct answer to the conditional question is no. But C will lie and say "yes," which means saying "ja."

Table 14.6.
The possible responses to Roberts's third question, under the assumption that C is not Random.

C's Identity	B's Identity	Da Means	C's Response
True	Random	Yes	Da
True	Random	No	Da
False	Random	Yes	Da
False	Random	No	Da
True	False	Yes	Ja
True	False	No	Ja
False	True	Yes	Ja
False	True	No	Ja

We see that a reply of da implies that C is True, while a reply of ja implies that C is False.

The final question is to ask C: "If I asked you if god B is Random, would you say 'da'?" The possible responses are recorded in Table 14.6. Since the reasoning here is so similar to what has come before, I will not trouble to provide any further explanation.

Once more we see a familiar pattern. A reply of "da" implies that B is Random. Since we already know C's identity, we can determine A's as well, and the problem would be solved. A reply of "ja" implies that B is not Random, which implies that A is Random. This, again, solves the problem.

To summarize, Roberts's solution is as follows:

- Direct the first question to B and ask: "If I asked you if god A was Random, would you say 'da'?" The reply allows you to know for certain either that A is not Random or that C is not Random. For simplicity, let us assume it is C who is not Random.
- Direct the second question to C and ask: "If I asked you if you always told the truth, would you say 'da'?" The reply will allow you to determine whether C is True or C is False.
- Direct the third question to C again and ask: "If I asked you if god B was Random, would you say 'da'?" From the reply you will determine either A's identity or B's identity, from which the identity of the third god immediately follows.

Well done! A bravura performance, and surely the last word on the subject. Roberts's solution is such a dramatic improvement over Boolos's that further improvements are unlikely.

Which is not to say, however, that further improvements are impossible.

14.3 Trivializing the Hardest Puzzle Ever

The next salvo was fired by Brian Rabern and Landon Rabern (2008). They noticed a a loophole in Boolos's presentation that permitted a very direct solution to the HLPE. In effect, you can apply the generalized Nelson Goodman principle as though the Random god were not even there.

The loophole is this: Random's behavior is not *entirely* random. According to Boolos, each time Random is asked a question, he flips a coin to determine whether he will behave as True or as False (this is explicit in the third bullet point in Section 14.1). The point, though, is that he is always behaving either as True or as False.

In the language of knight/knave puzzles, Boolos is explicit that Random behaves in the same manner as normals. Consequently, there are certain questions to which Random will respond in a predictable manner. An example of such a question is: "Have you ever uttered a lie?" True and False will both respond "no," meaning that Random will also respond "no," regardless of the result of his mental coin toss. Apparently, his answers are not always so random after all.

Here is another example, which Rabern and Rabern attribute to a C. Young from an unpublished manuscript. As in Section 14.1, let X denote the assertion: "Dushanbe is in Kirghizia." Now suppose we ask Random: "Is it true that you are lying if and only if X?" The analysis from previous sections is readily adapted to show that Random will answer negatively only when X is true, and affirmatively only when X is false.

The significance of this example is that our previous solutions labored under a false assumption. Specifically, it seemed as though no useful information could be gleaned from talking to Random, since his answers are impossible to interpret. Therefore, our first question had to be used to find a god who is definitely not Random. However, we can exploit the predictability of Random's responses to ask questions that are informative regardless of to whom they are addressed.

To show how, Rabern and Rabern establish a general principle. Let q be any yes/no question. Then we associate a new yes/no question, called $E(q)$, to it as follows:

$$E(q) = \text{``If I asked you } q \text{ in your current}$$

$$\text{mental state, would you say `ja'?''}$$

In other words, $E(q)$ is a function whose domain and codomain are the set of yes/no questions. Then regardless of the god to whom $E(q)$ is addressed, we can draw the following conclusions:

- A reply of "ja" indicates that the correct answer to q is affirmative.
- A reply of "da" indicates that the correct answer to q is negative.

Notice that the effect of $E(q)$ is to embed q within a larger question, which explains our use of the notation "E." Rabern and Rabern refer to these two bullet points as the Embedded Question Lemma (EQL).

It would be straightforward, but tedious, to make a table listing all the cases. The analysis proceeds in a manner similar to what we have discussed several times already. To get the flavor, let us consider the possibilities under the assumption that "ja" means yes. The analysis under the assumption that "ja" means no is very similar and will be left to you.

- If the correct answer to q is yes, then True will say "ja" when asked q. Therefore, the correct answer to $E(q)$ is also yes, so True's answer will again be "ja." In contrast, False will reply "da" when asked q. Thus, the correct answer to $E(q)$ is now no, but False will lie and say "ja" in response to $E(q)$.
- If the correct answer to q is no, then True will say "da" when asked q. Therefore, the correct answer to $E(q)$ is also no, and True will once again say "da." In contrast, False will say "ja" when asked q. Thus, the correct answer to $E(q)$ is now yes, but False will lie and say "da."

As noted by Rabern and Rabern, these cases essentially reduce to the fact that both a double negative and a double positive make a positive. In both cases, we hear "ja" when the correct response to q is affirmative, and "da" when the correct response is negative.

We have not yet considered how Random will respond. (We are continuing to assume that ja means yes.) Now, in anticipation of being asked a question, Random will have flipped his mental coin, and will therefore be either in the mental state of True or the mental state of False. Since we have already considered the behavior of both True and False, the case in which we are speaking to Random does not involve anything beyond what we have already considered.

Armed with the EQL, the Hardest Logic Puzzle Ever becomes trivial. The reason is that we now have a device for obtaining unambiguous answers to simple questions. As we did in previous sections, let us state this solution explicitly:

- Direct the first question to A and ask: "If I asked you if you are True in your current mental state, would you say 'ja'?" If he replies "ja," then we know he really *is* True. If he replies "da," then we know he is either False or Random.
- In the former case, we follow by asking B: "If I asked you if you are False in your current mental state, would you say 'ja'?" His answer will definitively identify him, from which C's identity immediately follows. In this case we are done in two questions.
- In the latter case, we ask A the follow-up: "If I asked you if you are False in your current mental state, would you say 'ja'?" Now his answer will definitively identify him. We will then only need a single question to identify B, after which we will be finished.

It is arguable, however, that this solution is a bit of a cheat. Rabern and Rabern were very clever to notice the imprecision of Boolos's phrasing, but their solution hardly gets to the heart of the puzzle. It seems clear that Boolos intended a truly random god, one who answers yes/no questions as though he did not even hear what was asked. We cannot really call it a day until we have solved this, almost certainly intended, version of the puzzle. Sensitive to this point, Rabern and Rabern show that even with a truly random god, the puzzle is not really so difficult.

In this version of the problem, the gods cannot change their mental state. For that reason, Rabern and Rabern suggest a modification to their function $E(q)$. If q is any yes/no question, we define the function $E^*(q)$ as follows:

$$E^*(q) = \text{If I asked you } q, \text{ would you say "ja"?}$$

We have already analyzed questions of this sort. In particular, we know that if $E^*(q)$ is directed to someone who is not Random, then a response of "ja" indicates that the correct answer to q is affirmative, while an answer of "da" indicates that the correct answer to q is negative.

However, with a truly random god, we must revert to the strategy employed by Boolos and Roberts. That is, the first question must be used to identify a god who is not Random. For that purpose, Rabern and Rabern define q to be the question: "Is A Random?" We then direct the question $E^*(q)$ to B. For reasons we have already discussed, if B says "ja," then we can be certain that C is not Random, and if B says "da," then we can be certain that A is not Random.

Assume that we have learned that A is not Random. Then the two further questions

1. E^*(Is A True?)
2. E^*(Is B Random?)

will finish off the problem.

Surely, that is the end of the discussion. We started with intricate, nested, biconditional questions. Now we have a device for eliciting direct answers to simple questions. Solving the HLPE is now seen to be trivial.

But have we really reached the simplest possible solution? Perhaps not, because we have not yet discussed whether three questions are really necessary.

14.4 Are Three Questions Necessary?

Let us return now to the original puzzle, as Boolos expressed it. We are no longer considering a truly random god, but rather one who, at any given moment, is in the mental state of either True or False.

At first glance, a two-question solution seems flatly impossible. This is because a yes/no question can only cut in half the number of possibilities that must be distinguished. For example, if I am thinking of a whole number between 1 and 8 inclusive, you might start by asking me if the number is among 1, 2, 3, and 4. However I respond, the number of possibilities will have been reduced from eight to four. If we assume the answer is yes, then your next question will be whether the number is among 1 and 2. My answer will reduce the number of possibilities from four to two. If the answer is no this time, your final question will be whether you are thinking of 3. Regardless of how I answer, you will then know my number.

The general principle is that if I start with n possibilities, I will need $\log_2(n)$ yes/no questions to narrow things to one possibility, where it is understood that I round up to the next whole number if the logarithm evaluates to a fraction. In the HLPE, the three gods can be arranged in six ways. That is too many to be reduced to one with two yes/no questions.

That seems convincing, but Rabern and Rabern note a possible flaw. You see, there are devious questions that certain gods cannot answer while remaining true to their natures. They give the example: "Are you going to answer 'ja' to this question?" They write:

> If 'ja' means *no*, then True will be unable to respond with the truth. If 'ja' means *yes*, then False will be unable to respond with a lie. But they are infallible gods! They have but one recourse—their heads explode. (Rabern and Rabern 2008, 109)

The gods might instead choose to smite the unwise questioner who taunted them in this way. That aside, their suggestion opens an intriguing possibility. If we construe "exploding head" as a third answer to a yes/no question, then we can distinguish three possibilities with one question. A two-question solution might be possible after all.

Let us examine this possibility more closely. Suppose we ask one of the gods: "Are you going to answer this question with the word that means 'no' in your language?" Random will just make up an answer and experience no cognitive dissonance. False will just answer with the word that means "yes," and he also will experience no dissonance. But what can True do? If he answers with the word that means "yes," then his answer is untrue, which is contrary to his nature. If he answers with the word that means "no," then he has again told an untruth, since the correct answer in this case is the word that means "yes." Consequently, his head will explode.

Just think of the consequences! We ask this question of one of the gods. If his head explodes, then he is True. If his head does not explode, then we ask the follow-up question: "Are you going to answer this question with the word that means 'yes' in your language?" This time the analysis shows that False cannot answer the question. So, if the god's head explodes, then we know he is False. If his head still has not exploded, then we have identified him as Random. This shows that the identity of any god can be discovered in no more than two questions.

This is clever, but can we really use the possibility of an exploding head to devise a two-question solution to the HLPE? Rabern and Rabern show that we can, but it will take a very clever question indeed. They suggest directing the following question to A. We refer to this question as QUERY–1:

> E(Is it the case that: [In your current mental state, you would always answer "da" to QUERY–1 and (B is true)] or (B is False)?)

It will take some work to parse this and to work out all the cases.

First, this is an instance of an embedded question in the sense of the EQL. Taking q to be the question enclosed by the outermost parentheses, we know that an answer of "ja" indicates that the correct answer to q is affirmative, while an answer of "da" indicates that the correct answer to q is negative.

Now let us take a closer look at q. It is the question:

> Is it the case that: [In your current mental state, you would always answer "da" to QUERY–1 and (B is True)] or (B is False)?

The question consists of two clauses joined with "or." These two clauses are:

1. (In your current mental state, you would always answer "da" to QUERY–1) and (B is True).
2. B is False.

There are three cases to consider, depending on how A answers. For each case, we will show that B's identity is determined.

- Suppose A answers "ja." By the EQL, this implies that the correct answer to q is affirmative. If B is not False, then clause 2 is false. This would imply that clause 1 is true. But since clause 1 is a conjunction, it is true only when both parts of it are true. But the first part requires that A would always answer "da" to QUERY–1. This is false, since in fact A just answered "ja" to QUERY–1. We have reached a contradiction. This shows that if A answers "ja," then B is False.
- Suppose A answers "da." Again invoking the EQL, we conclude that the correct answer to q is negative. This would imply that clauses 1 and 2 are both false in this case. Since clause 2 is false, we know that B is not False. And since clause 1 must be false despite A having answered "da" to QUERY–1, we see that B is not True either. This shows that if A answers "da," then B is Random.
- Finally, suppose that A's head explodes. If B is Random, then clauses 1 and 2 are both false. Thus the correct answer to q is negative. In this case, A would simply say "da," and his head would not explode. If B is False, then clause 2 is true. But then the correct answer to q is affirmative. God A would simply say "ja," and again his head would not explode. This shows that if A's head explodes, then it could only be because B is True.

Just like that, we have determined B's identity in one question. Now that we have accomplished this, the second question can employ the EQL to determine the identities of the remaining gods with one further question. For example, we can pick any identity that B is not, and then ask B: "If I asked you if C was this god, would you say 'ja'?" The answer to this will definitely identify C and therefore A as well.

It looks like a two-question solution to the HLPE is possible after all, at least if we can countenance a creative interpretation of the rules.

We could end the story here, but that would leave a big loose end. Rabern and Rabern solved Boolos's original puzzle in two questions, but what about the version with a truly random god? Can *that* version be solved in just two questions?

14.5 Two Questions When Random Is Really Random

Gabriel Uzquiano (2010) showed that even with a truly random god, the HLPE can *still* be solved with just two questions. His approach, however, exploited a different loophole from what we have seen for extracting three answers from a binary question. Whereas Rabern and Rabern used the fact that some self-referential questions cannot be answered by certain gods in a manner consistent with their natures, Uzquiano used the fact that some questions cannot be answered by certain gods, because their answers are simply unknown.

In particular, what could True and False possibly say in response to a query about how Random would answer a question? Since we are now assuming that Random just invents an answer on the spur of the moment, True and False would have no way of knowing what he would say. This will not cause their heads to explode, since they are not forced into contradiction, but it *will* force them to remain silent. Thus, silence is our third possible response to a binary question.

By asking suitably ingenious questions, we can use this fact to solve the HLPE in two questions, even assuming a truly random god. Since Uzquiano's two questions are rather complex, we will build up to them by solving a simpler puzzle first.

Puzzle 87. *As before, we have the three gods True, False, and Random. True always speaks truthfully, False always speaks falsely. Random answers entirely at random; he just makes up an answer to your question as though he did not even hear what you said. Moreover, all three gods will respond in English to any yes/no question put to them, always in a manner consistent with their natures. (We assume that Random will respond with either "yes" or "no," choosing which one of them to say in some random manner.) What should we ask to determine the identities of the gods in the smallest number of questions?*

Uzquiano solves this puzzle using two questions. Inspired by Boolos, he suggests that we begin by asking A: "Would you and B give the same answer to the question of whether Dushanbe is in Kirghizia?" There are three cases, depending on how A responds:

1. If A cannot answer the question, we know immediately that B is Random. In this case, A is either True or False.
2. Suppose A answers "yes." Then we have two subcases to consider:
   • A is Random and has answered randomly.
   • A is not Random. Since A answered, we know B is not Random. It follows that A and B are True and False in some order, implying they would answer the question of Dushanbe's location differently. Thus, A's answer of "yes" is false, implying that A is False and B is True.
3. Suppose A answers "no." Then again we have two subcases:
   • A is Random and has answered randomly.

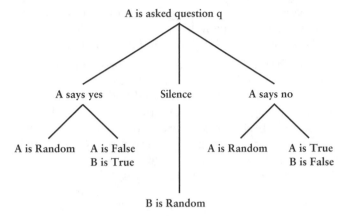

Figure 14.1. The possible scenarios when A is asked the question (denoted by q): "Would you and B give the same answer to the question of whether Dushanbe is in Kirghizia?" Note that if A responds with "yes" or "no," then we know that B is not Random. If A remains silent, then we know that B *is* Random.

- A is not Random. In this case, an analysis that is nearly identical to the previous case shows that A is True and B is False.

The various scenarios are shown in Figure 14.1.

The upshot is that regardless of how A answers, we will once more identify a god who is not Random. If A is silent, then we can be certain that B is Random (and therefore neither A nor C is Random); and if A answers, then we can be sure that B is not Random.

We can now direct our second question to this god, who is either A or B, and ask: "Would you and C give the same answer to the question of whether Dushanbe is in Kirghizia?" We again have three cases, depending on the response:

- If there is no response, then it must be that C is Random. Now that we know that C is Random, we can use A's response to the first question to determine whether he is True or False. Specifically, if he answered "yes," then he is False, and if he answered "no," then he is True, as shown in Figure 14.1 (keeping in mind that the branches in which A is Random are no longer live options.)
- If the answer is "yes," then C cannot be Random. In this case, the correct answer to the question is "no," implying that we directed the question to False. It follows that C is True, and the remaining god is Random.

- If the answer is "no," then C again cannot be Random. In this case, the god we are speaking to has answered truthfully, and is therefore True. This implies that C is False, and the remaining god is Random.

This is not yet a solution to the HLPE, of course, since we have assumed that the gods respond in English. We need to modify our questions so that they continue to work even if they respond with "da" and "ja."

At this point, though, we know that this poses no problem at all! The EQL from Section 14.4 shows us precisely what to do. Recall that the EQL tells us that if we ask any god who is not really Random: "If I asked you q in your current mental state, would you say 'ja'?" then an answer of "da" indicates that the correct answer to q is affirmative, and an answer of "ja" indicates that the correct answer to q is negative.

Uzquiano uses this idea, with a slightly different phrasing, to adapt his solution to the HLPE. We will omit the analysis of the specific cases, since we have enough experience at this point to know roughly how they will play out. Here, then, is Uzquiano's two-question solution to the HLPE under the assumption of a truly random god:

- Direct to A the question: "Would you answer 'ja' to the question of whether you would answer with a word that means yes in your language to the question of whether you and B would give the same answer to the question of whether Dushanbe is in Kirghizia?" The answer to this question will identify either A or B as not being Random. For simplicity, assume that it is B who is not Random.
- Direct to B the question: "Would you answer 'ja' to the question of whether you would answer with a word that means yes in your language to the question of whether you and C would give the same answer to the question of whether Dushanbe is in Kirghizia? The answer will definitively identify B, and therefore A and C as well.

Very clever, as always, but arguably unsatisfying. After all, this solution depends on the idea that True and False must simply stand mute when asked about how Random will respond to a question. The problem is that all statements of the HLPE are explicit that we are dealing with gods. Typically, gods are thought to be omniscient. We might therefore assume that True and False know how Random will answer even before Random himself knows what he will say. With this assumption, our ingenious two-question solution is out the window.

Uzquiano anticipated this objection and goes on to show that we can still solve the HLPE in two questions. However, this will require being even more ingenious with our questions than we have been to this point.

To avoid belaboring things unnecessarily, let us proceed directly to the answer. We begin by asking A the following question, the entirety of which shall be referred to as QUESTION 1:

Would you answer "ja" to the question of whether either

(a) B is not Random and you are False, or

(b) B is Random and you would answer "da" to QUESTION 1?

The analysis of the specific cases is sufficiently similar to what we have already seen that I think it would be more tedious than helpful to work through everything again. Suffice it to say that the response to this question will definitively identify one of A or B as being not Random. We then direct the following question to that god, the entirety of which shall be referred to as QUESTION 2:

Would you answer "ja" to the question of whether either

(a) C is not Random and you are False, or

(b) C is Random and you would answer "da" to QUESTION 2?

The answer to this second question will reveal the identities of the three gods.

Moreover, as promised, this solution assumes that True and False somehow know what Random's answers to questions will be, even before he formulates them himself.

But even this is not yet the whole story, for we might still feel that Random is being treated unfairly. Consider the evolution in our understanding of Random's behavior. Boolos's original idea was that Random was something of a chameleon, sometimes analyzing situations as True analyzes them, and sometimes analyzing situations as False analyzes them. Rabern and Rabern showed that this conception of Random trivializes the problem, and recommended replacing him with a god who truly answers randomly. So, he is like True and False in the sense that he will always answer with "da" or "ja" to any yes/no question, but his answer has no connection to the question he was asked. But then Uzquiano pointed out that True and False might have the option of not answering at all and that we can use this fact to our advantage.

It would seem that if we *really* want Random to be like True and False, we must also give *him* the option of remaining silent in response to a question. Given this option, the solutions considered in this section no longer work.

So how many questions do we need now?

14.6 *What If Random Can Remain Silent?*

It turns out that the answer to that question is: It depends.

According to Gregory Wheeler and Pedro Barahona (2012), the answer is three. Why would they say this?

In Section 14.4, we argued that three questions were necessary for solving Boolos's version of the HLPE. Our reasoning was simple: A binary question can only divide a sample space in half. This implies that two binary questions can only distinguish among four possibilities. But the three gods in the HLPE can arrange themselves in six different ways. That is too many to handle with just two questions, implying that three questions are necessary.

That is a cogent argument, but first Rabern and Rabern, and then Uzquiano, found ways around it. They did this by noting that there were questions that could not be answered by certain gods in a manner consistent with their nature, either because these questions led to contradictions or because the gods lacked knowledge of Random's future behavior.

This was important, because it created an asymmetry in the behaviors of True and False on one hand and of Random on the other. Specifically, whereas Random always gave a verbal answer to any question put to him, True and False did not. When speaking to True and False, there was the possibility that their heads would explode (in the Rabern and Rabern version) or that they would remain silent (in Uzquiano's version), but neither of these occurrences is possible when speaking to Random. A sufficiently clever question could take advantage of this asymmetry, to make certain possible answers to the opening question inconsistent with the premise that you were speaking to Random.

More precisely, it is a general principle that a question with n possible responses cannot reduce a sample space to a single possibility if the space begins with more than n elements. However, a successful two-question solution to the HLPE must reduce six possibilities to one using questions with three possible answers. This is possible only if the first question can reduce the space to no more than three possibilities. Otherwise, the second question will have to distinguish among four possibilities, which is not possible when there are only three possible responses. The logic is nicely illustrated by the tree diagram in Figure 14.1. If you trace through the three branches corresponding to the answers "yes," "no," and silence, then you will find that each of A's possible responses left no more than three active possibilities.

It is precisely this asymmetry that is eliminated in Uzquiano's version of the problem. If Random is allowed to remain silent, then there is simply no response to your opening question that is inconsistent with the premise that you are speaking to Random. If the response is "da" or "ja," it might be that you are speaking to True or False and that their answer was the result of some process of ratiocination, or it could be Random just making something up. If the response is silence, that might be because True or False found themselves unable to answer, or it might be that Random just chose to remain silent.

Let us assume you direct your first question to A. Then regardless of how A answers, the two permutations in which A is Random (A = Random, B = True, C = False, and A = Random, B = False, C = True) are guaranteed to be live possibilities regardless of the response you receive. This puts a limit on how effective your first question can be. The three possible responses must distinguish among the other four possibilities. It follows that there must be at least one response after which two of those four possibilities remain viable. Since we know the two scenarios in which A is Random will remain viable, we conclude that there is at least one response that leaves four viable possibilities. That is too many possibilities.

And that is why three questions are necessary if Random is allowed to remain silent at his own whim.

Compelling stuff. Absolutely convincing. We are *done*.

Maybe not! Shortly after the publication of the Wheeler and Barahona (2012) paper, philosopher Stefan Wintein noted that their argument depends critically on the assumption that the gods have only three possible responses to the questions that are put to them (Wintein 2012). That assumption, in turn, is based on a certain conception of what it means to answer truthfully or falsely to a binary question. Wheeler and Barahona took it for granted that the gods remained silent in the face of questions that led to inconsistency or required knowledge of future events. This is not the only assumption that might have been made.

We have already seen that self-referential questions can provide challenges to gods who must answer either truthfully or falsely. For example, let us refer to the following question as λ:

Is it the case that your answer to λ is no?

If you answer this question with either "yes" or "no," you will have answered falsely. In Wheeler and Barahona's conception of the puzzle, True will have to remain silent when confronted with this question.

Providing the flip side to this observation are questions like the following, which we denote by τ:

Is it the case that your answer to τ is yes?

This time, regardless of whether you answer with "yes" or "no," you will have spoken truthfully. Previous discussions of the HLPE gave no consideration to questions of this sort, and it is not clear how those previous authors would have handled them.

Wintein's solution to this dilemma is to allow "Both" and "Neither" as possible answers the gods might give to the binary questions they are asked. Explicitly, True would say:

The question λ can be answered with *neither* "yes" nor "no." The question τ can be answered with *both* yes and no.

For False, the situation is reversed. False would say:

> The question λ can be answered with *both* yes and no. The question τ can be answered with *neither* "yes" nor "no."

Now that we have four possible responses to our questions, it might be possible after all to solve Uzquiano's puzzle in just two questions.

In fact, Wintein proposes just such a solution. He considers first the case where we assume the gods respond in English, and we will follow his example. His solution begins by posing the following question, which he refers to as α_1:

> Is it the case that (you will answer "no" to α_1 and A is Random) or (you will answer "yes" to α_1 and B is Random) or (C is Random)?

To make the analysis simpler, note that α_1 is a disjunction with three parts:

1. You will answer "no" to α_1 and A is Random.
2. You will answer "yes" to α_1 and B is Random.
3. C is Random.

Recall that a disjunction is true if any one of its parts is true, and false only if all of its parts are false.

Since the analysis includes elements we have not previously seen, let us work through all the cases. Our strategy will be to ask how True and False would respond to this question under different assumptions about which god is Random. There are six cases to consider: Any of the three gods could be Random, and α_1 could be answered with either "yes" or "no." It will be convenient, in what follows, to refer casually to the truth or falsity of α_1 (by this I mean the truth of the underlying disjunction). The findings of this analysis are summarized in Table 14.7.

1. Suppose A is Random.
   - If α_1 is answered "yes," then all three parts of the disjunction are false, implying that α_1 is false. Thus, it is incorrect to answer "yes" when A is Random.
   - If α_1 is answered "no," then part 1 of the disjunction is true, implying that α_1 is true. Thus, it is incorrect to answer "no" when A is Random.

   It follows that True will respond with "neither" to α_1, while False will respond with "both."
2. Suppose B is Random.
   - If α_1 is answered "yes," then part 2 of the disjunction is true, implying that α_1 is true as well. Thus, it is correct to answer "yes" when B is Random.

TABLE 14.7.
The responses of True and False to the question α_1.

Random?	Yes or No?	$V(\alpha_1)$	✓ or ×	True	False
A is Random	Yes	False	×	Neither	Both
	No	True	×		
B is Random	Yes	True	✓	Both	Neither
	No	False	✓		
C is Random	Yes	True	✓	Yes	No
	No	True	×		

Note: The first two columns represent assumptions we might make about who is Random and the response we receive to α_1. $V(\alpha_1)$ denotes the truth value of α_1, under the assumptions made in the first two columns. The symbols ✓ and × denote that the given response to α_1 is, respectively, correct or incorrect. The final two columns show how True and False would respond.

- If α_1 is answered "no," then all three parts of the disjunction are false, implying that α_1 is false. Thus, it is correct to answer "no" when B is Random.

  It follows that, this time, True will respond with "both" to α_1, while False will respond with "neither."

3. Suppose C is Random. In this case, part 3 is true regardless of whether we answer yes or no, implying that α_1 is true as well. Thus, it is correct to answer "yes" and incorrect to answer "no" when C is Random. Therefore, True will answer "yes," and False will answer "no."

Consideration of Table 14.7 now reveals that an answer of "yes" or "no" implies that C is Random (and therefore that B is not), and a reply of neither or both implies that either A or B is Random (and therefore that C is not). Thus, this first question definitively identifies a god who is not Random.

For the second question, let us assume that we have identified B as not Random. Then there are four live scenarios:

$$p_1 = \text{B is True, A is False, C is Random}$$

$$p_2 = \text{B is True, A is Random, C is False}$$

$$p_3 = \text{B is False, A is True, C is Random}$$

$$p_4 = \text{B is False, A is Random, C is True}$$

TABLE 14.8.
The responses of True and False to the question α_2.

Scenario	Yes or No?	$V(\alpha_1)$	✓ or ×	True	False
p_1	Yes	False	×	Neither	Both
	No	True	×		
p_2	Yes	True	✓	Both	Neither
	No	False	✓		
p_3	Yes	True	✓	Yes	No
	No	True	×		
p_4	Yes	False	×	No	Yes
	No	False	✓		

Note: The notation is the same as that defined in Table 14.7.

Continue to assume that our first question revealed that B is not Random. Then we ask him the following question, which we refer to as α_2:

> Is it the case that (you will answer "no" to α_2 and p_1 is correct) or (you will answer "yes" to α_2 and p_2 is correct) or (p_3 is correct)?

The analysis plays out in a manner similar to before. Notice that our assessments of what is true and false, and correct and incorrect, depend in part on which of the four scenarios $p_1, \ldots, p_4$ we are in. The results are presented in Table 14.8.

The manner in which the responses to α_2 reveal the identities of the gods is truly beautiful. Suppose B responds "yes" to α_2. Then we see from Table 14.8 that there are two possibilities: B is True and we are in p_3, or B is False and we are in p_4. But in scenario p_3, B is False. This shows that if B responds "yes," then scenario p_4 is correct.

What if B responds "neither"? Then the two possibilities are: B is True and we are in p_1, or B is False and we are in p_2. But since B is True in scenario p_2, we know that scenario p_1 must be correct. Similar reasoning quickly shows that a response of "both" implies that p_2 is correct, while a response of "no" implies that p_3 is correct. Absolutely brilliant.

Of course, this strategy only solves the problem when the gods are speaking English. Modifications will be needed to handle the complications of "da" and "ja." I will show you the questions that get the job done, just for the sake of completeness. However, the analysis of the cases is very tedious and is similar to what we have already seen. So, you are on your own if you want to understand why these questions solve the problem. Good luck!

Here are the two questions. First, ask A question β_1, which is the following:

> Is it the case that da means "yes" if and only if ((you will answer "da" to β_1 and A is Random) or (you will answer "ja' to β_1 and B is Random) or (C is Random))?

The response to this question will definitely identify one of the gods as not Random. Let us assume once more that it is god B. Moreover, we refer to the same scenarios $p_1, \ldots, p_4$ as before. We then ask B question β_2, which is the following:

> Is it the case that da means "yes" if and only if ((you will answer "da" to β_2 and p_1 is correct) or (you will answer "ja" to β_2 and p_2 is correct) or (p_3 is correct))?

I spent the better part of an afternoon working through all the cases to make sure everything worked out the way Wintein said. They really do! Frankly, though, I think these questions are beautiful to look at just as though they were paintings. With the reappearance of biconditional questions, they even bring us full circle back to Boolos's original solution.

A perfect place to end our account of the saga of the Hardest Logic Puzzle Ever!

14.7 Notes and Further Reading

Of course, in philosophy, nothing is ever the last word. There continues to be work done on the Hardest Logic Puzzle Ever. Having read to this point in the book, you might be wondering what would happen if the gods chose to use a nonclassical logic. A first step toward answering that question is provided in a recent paper by Carnielli (2017). I have chosen to omit a discussion of this paper here, on the grounds that enough is enough.

CHAPTER 15

Metapuzzles

Logic puzzles are all about determining the deductive consequences of given information. In Lewis Carroll's puzzles, the given information was presented in the form of humorous categorical propositions. Raymond Smullyan countenanced the more general sorts of propositions uttered by knights and knaves.

We could imagine another sort of puzzle, however, in which the given information comes in the form of statements made about what other people were able to infer from information *they* had received. Smullyan referred to such puzzles as "metapuzzles," by which he meant puzzles that could only be solved by knowing about other puzzles that could or could not be solved. He was especially adept at devising such puzzles, as we shall see.

Smullyan did not invent the genre, so the present chapter will discuss a few classics as well. These puzzles can be surprisingly difficult, so some patience is in order as you work your way through them. The solution to the puzzle presented in Section 15.1 will be presented in the main text, to illustrate possible strategies for solving. Solutions to all other puzzles will appear at the end of the chapter.

15.1 A Warm-Up Puzzle

We open with a straightforward representative of the genre. In what follows, you should assume that Arne is a perfect logician. If he has enough information to solve a problem, then he will, indeed, solve the problem. You should make similar assumptions for all the puzzles in this chapter. Puzzle 88 is a slightly modified version of a puzzle presented by Gardner (2006, 372–373).

Puzzle 88. *Arne and Beth were having a conversation. Arne said, "What are the ages, in years only, of your three children?"*

Beth replied, "The product of their ages is 36."

Arne shook his head. "That's not enough information," he said.

"The sum of their ages is the same as your son's age," said Beth.

"That's still not enough information."

Beth paused. Then she said, "My oldest child—and he's at least a year older than either of the others—has a wart on his left thumb."

At this point Arne was able to deduce the ages of the children. Can you do likewise?

How might we solve such a thing? A natural place to start is with the fact that their ages multiply out to 36. There are only eight ways of expressing 36 as the product of three positive integers:

$$1, 1, 36 \quad 1, 2, 18 \quad 1, 3, 12 \quad 1, 4, 9$$
$$1, 6, 6 \quad 2, 2, 9 \quad 2, 3, 6 \quad 3, 4, 3.$$

If we knew the age of Arne's son, then we might be able to get somewhere. Since we do not, we seem to be stuck.

However, Arne certainly knows his son's age, and we know this was insufficient for him to narrow the eight possibilities down to one. Most of the triples on the list have a unique sum. For example, $1, 1, 36$ is the only triple that sums to 38, and $1, 2, 18$ is the only triple that sums to 21. If the age of Arne's son was 38 or 21, then he would not have had to ask anything further to deduce the ages of Beth's children. Since he did need more information, we conclude that the correct triple does not give a unique sum. The only triples that give the same sum are $1, 6, 6$ and $2, 2, 9$. We have thus narrowed the eight possibilities down to two.

Arne then learns that Beth has a uniquely oldest child, and that eliminates $1, 6, 6$. It follows that the correct answer is $2, 2, 9$.

Very nice, and a helpful template for what is to come.

This might be a good time to mention that when solving these puzzles, we worry only about mathematical possibilities. We do not ask questions like: "Who has 1-year-old twins when they already have a 36-year-old child?"

15.2 The Playful Children and Caliban's Will

The warm-up puzzle is actually a simplified version of one of the real classics of the genre. It has been anthologized in many puzzle books, but it was originally presented by L. R. Ford in 1947. To solve his presentation of the puzzle, you will need to know that a baseball team has nine players.

Puzzle 89. *"Are those your children that I hear playing in the garden?"* *asked the visitor.*

"There are really four families of children," replied the host. "Mine is the largest, my brother's family is smaller, my sister's family is smaller still, and my cousin's is the smallest of all. They are playing drop the handkerchief," he went on; "they prefer baseball but there are not enough children to make two teams. Curiously enough," he mused, "the product of the numbers in the four groups is my house number, which you saw when you came in."

"I am something of a mathematician," said the visitor, "let me see whether I can find the numbers of children in the various families. After figuring for a time he said, "I need more information. Does your cousin's family consist of a single child?" The host answered his question, whereupon the visitor said, "Knowing your house number and knowing the answer to my question, I can now deduce the exact number of children in each family."

How many children were there in each of the four families? (Ford 1947, 339–340)

Another classic is the puzzle of Caliban's will, first published in a newspaper column run by Hubert Phillips in 1933. The somewhat eccentric names of the characters refer to some of Phillips's actual colleagues. The puzzle itself was devised by Max Newmann (Bellos 2017, 28).

Puzzle 90. *When Caliban's will was opened, it was found to contain the following clause:*

> *I will leave ten of my books to each of Low, Y. Y., and 'Critic,' who are to choose in a certain order:*
> * *No person who has seen me in a green tie is to choose before Low.*
> * *If Y. Y. was not in Oxford in 1920, the first chooser never lent me an umbrella.*
> * *If Y. Y. or 'Critic' has second choice, 'Critic' comes before the one who first fell in love.*

Unfortunately, Low, Y. Y., and 'Critic' could not remember any of the relevant facts; but the family solicitor pointed out that, assuming the problem to be properly constructed (i.e., assuming it to contain no statement superfluous to its solution) the relevant data and order could be inferred. So, what was the prescribed order of choosing? (Bellos 2017, 29)

The key to the puzzle is that there is no superfluous information in the three given premises. So, for example, we know immediately that Low is not the second chooser, for if he were, then the third statement would provide no information. There is still a lot more work to do after noticing that, so get started!

15.3 Knight/Knave Metapuzzles

Smullyan was especially fond of metapuzzles and published many of them in his various books. I have chosen two as representative, but be warned that they are quite difficult.

Puzzle 91. *A logician visited a knight/knave island and met two inhabitants, A and B. He asked A, "Is it true that B once said that you are a knave?" A answered yes or no, but we do not know which. Then the logician asked one of the two whether the other was a knave. He received a yes/no answer. It is not given whether the logician was now able to determine the types of A and B. Later a second logician met the same pair. He asked A whether B had once claimed that A and B were both knaves. A answered (yes or no). Then the logician asked one of them whether the other was a knave and was answered (yes or no). Again, it is not given whether the logician could now determine the types of A and B. However, it is known that one logician was able to solve the problem and the other was not. Can you now determine the types of A and B? (Smullyan 1992a, 94–95)*

Puzzle 92. *At a certain trial, the prosecutor said, "The defendant is guilty and has committed other crimes in the past." The defense attorney replied, "The defendant is innocent and the prosecutor is a knave." The judge then asked a question whose answer he already knew. One of the attorneys answered it, thus revealing to the judge whether that attorney was a knight or a knave. From this, the judge could determine whether the defendant was guilty.*

A reporter present at the trial described all of this to a logician. The logician found he had insufficient information to determine whether the defendant was guilty. He asked the reporter, "If the other attorney had answered the question instead, would the judge then have known whether the defendant was guilty?" The reporter thought for a while and replied, "I really don't know. In fact, there is no way of knowing." The logician replied, "Oh, good. Now I know whether the defendant was innocent or guilty."

Can you determine whether the defendant was innocent or guilty? (Smullyan 1992a, 95–96)

15.4 Solutions

Puzzle 89. The visitor is initially given two pieces of information: The numbers of children are distinct, and their sum is less than 18. If the smallest family had 3 children, then the smallest total number of children occurs when there are 3, 4, 5, and 6 children in the families. This makes 18, which is too many. It follows that the cousin's family, which is identified in the puzzle as the smallest, has either 1 child or 2 children.

For the sake of argument, let us suppose that the smallest family has 2 children. Then we can easily list all of the possible sequences of four numbers that could represent the distribution of numbers of children:

$$2345 \quad 2346 \quad 2347 \quad 2356 \quad 2348 \quad 2357 \quad 2456.$$

We are also given that the visitor knows the host's house number, and that this house number is equal to the product of the numbers of children. If the cousin has only 1 child, then we can check that the largest product occurs from the sequence 1358, which multiplies out to 120. In our list, every sequence other than 2345 gives a product that is larger than 120.

From this we conclude that if the house number is larger than 120, then the visitor would know the cousin had 2 children, while if the house number is smaller than 120, then he would know the cousin had 1 child. Either way, he would not have needed to ask how many children the cousin had. Since he did ask, we see that the house number must be 120.

The visitor then asked if the cousin had a single child. Had the answer been yes, he would not have been able to distinguish between the sequences 1358 and 1456. The information he had would have been consistent with either possibility. Since the host's response entirely settled the question for the visitor, we conclude that the answer was no, and that the correct sequence is 2345.

Puzzle 90. Recall that we are assuming that all three bullet points contain relevant information. Thus, it cannot be that neither Y. Y. nor 'Critic' had ever seen Caliban in a green tie, because if this were the case, then the first bullet point tells us nothing. The first bullet point also entails that Low does not choose last.

From the second bullet point, we conclude that Y. Y. was not in Oxford in 1920, for if he were, the point would be uninformative. For the same reason, we can assume that *someone* lent Caliban an umbrella. But who? If Low lent the umbrella, then we would know he must be second. But if Low is second, then the third statement is uninformative. It follows that Low did not lend Caliban an umbrella.

Could it be that both Y. Y. and 'Critic' lent umbrellas to Caliban? The answer is no. If they both lent umbrellas, then it would follow immediately that Low, as the only one not to lend an umbrella, must be the first chooser. But then, for the third point to be informative, it must be that 'Critic' is second and Y. Y. is third. In this case, we will have determined the order with no reference to item one, which renders this solution impossible.

Who saw Caliban in a green tie? If both Y. Y. and 'Critic' saw Caliban in a green tie, then it would follow from the first point that Low chooses first. But this would render the second point uninformative.

To sum up the last two paragraphs: Either Y. Y. or 'Critic' (but not both) lent Caliban an umbrella, and either Y. Y. or 'Critic' (but not both) saw Caliban in a green tie.

There is one more case we must eliminate. Could it be that Y. Y. is the one who saw Caliban in a green tie and who also lent Caliban an umbrella? This is impossible, for the following reason: From the first point, we would know that Y. Y. cannot be first. In this case, however, the second point would be uninformative.

It follows that if Y. Y. saw Caliban in a green tie, then he is not the one who lent Caliban an umbrella. This would imply that 'Critic' lent the umbrella. By similar reasoning, if 'Critic' saw Caliban in a green tie, then Y. Y. must have lent him the umbrella.

In either scenario, it follows that Low must be first. (The first point implies that Low is not the last chooser, and the third point is only informative if either Y. Y. or 'Critic' has the second choice.) It follows that either Y. Y. or 'Critic' has second choice. The third point now implies that 'Critic' must be second. Thus, the final order is Low, 'Critic,' Y. Y.

Puzzle 91. We begin with the first encounter.

Person A answered either "yes" or "no" to the first logician's first question:

- If he answered "yes," then A affirmed that B once called him a knave. If A is a knight, then B really did say that A is a knave, and therefore B is a knave. It follows that A and B are not both knights.
- If A answered "no," Then A denies that B once called him a knave. If A is a knave, then B actually did once call him a knave, implying that B is a knight. It follows that A and B are not both knaves.

The second question reveals whether or not A and B are the same type. A straightforward argument shows that if either affirms that the other is a knave, then the two must be of different types, while if either denies the other is a knave, then they must be the same type. Now, from the first answer, the logician knows either that A and B are not both knights or that A and B are not both knaves. If the answer to the second question revealed that A and B were the same type, then he would be able to solve the problem. If the answer to the second question revealed that A and B were not of the same type, then he would be unable to solve the problem.

We now consider the second encounter. As before, A answered either "yes" or "no" to the second logician's first question.

- Suppose A affirmed that B once claimed they were both knaves. If A is a knight, then B must be a knave. Thus, they are not both knights.
- Suppose A denied that B once claimed they were both knaves. Then A cannot be a knave, for then the truth is that B really did claim they were both knaves, which leads to a contradiction, regardless of which type you assume B to be. So A must be a knight, and B never made such a claim. Nothing would follow about B, and the logician would know only that A is a knight.

Now consider the second question. As before, a "yes" answer to the second question shows A and B to be of different types, while a "no" answer shows them to be of the same type. If the logician learned they are of the same type, then he would have been able to solve the problem. Because if they are the same type, but not both knights, then both are knaves, and if they are the same type and A is a knight, then they are both knights. Suppose instead the logician learns that they are different. Then in our first scenario, the problem would not be solvable, while in the second scenario, the logician would learn that A is a knight and B is a knave.

Thus, if A and B are of the same type, then both logicians could solve the problem, which contradicts what we are told. It follows that A and B are of different types. Then the first logician could not solve the problem, while the second logician could only solve the problem in the scenario where A is a knight and B is a knave, and that is the answer.

Puzzle 92. One of the attorneys answered the question, and thereby revealed himself to be a knight or a knave. So we must consider four cases:

- CASE ONE. *The prosecutor revealed himself to be a knight.* In this case, the judge knew the defendant was guilty (and that the defense attorney was a knave).
- CASE TWO. *The prosecutor revealed that he was a knave.* Then the judge could not know whether the defendant was guilty. He could be innocent. He could also be guilty, but, contrary to the prosecutor's claim, he might never have committed any crimes in the past.
- CASE THREE. *The defense attorney revealed that he was a knight.* Then the judge would know the defendant was innocent (and that the prosecutor was a knave).
- CASE FOUR. *The defense attorney revealed himself to be a knave.* In this case, the defense attorney's statement is false. Thus, it cannot

be that the defendant is innocent *and* that the prosecutor is a knave. Hence, either it is false that the defendant is innocent or it is false that the prosecutor is a knave. In the former case, the defendant is guilty. In the latter, the prosecutor is a knight, in which case the defendant is again guilty. Either way, the judge would now know the defendant was guilty.

We next consider the reporter's reaction to the logician's question. Since the reporter was present at the trial, he knew which attorney answered the judge's question. Had it been the defense attorney who answered, then, from our analysis of cases three and four, the judge would have been able to determine the defendant's innocence or guilt. Thus, if the prosecutor had answered, the reporter would have answered "yes" to the logician's question. Since he did not, we see that the defense attorney answered the question and thereby revealed his type to the judge.

If the defense attorney is a knight, then the prosecutor is a knave. Therefore, had he, and not the defense attorney, answered the question, the judge could not have determined the defendant's guilt or innocence (as seen from case two). The reporter would then have answered "no" to the logician's question. Since he did not, the defense attorney must have revealed himself to be a knave. It is only in this case that the reporter would not have known how to answer the logician's question.

Since the defense attorney is a knave, we see from case four that the defendant is guilty.

CHAPTER 16

Paradoxes

My introduction to paradox came when I was 5 years old. My father asked me, "If you are standing on the Earth, and you want to see the Moon, what do you do?" I replied that you should look up. "Very good," my father said. "Now, if you are standing on the Moon and you want to see the Earth, what do you do?" Of course, I replied that you should look down. Then my father sprang his trap. He said, "But Jason, if you look down, you will see your feet."

I am older now, and I have the benefit of a serious science education to help me answer such questions. For all of that, however, it still seems strange to me that regardless of whether you are standing on the Earth or the Moon, you must look up to see the other.

I also recall the time in college when one of my mathematics professors explained that he was going to prove that all whole numbers are interesting. He asked us to suppose that there was at least one uninteresting whole number. Well, if there is at least one such number, then it follows that there must be a least such number. Call that number x. Then x is the smallest uninteresting whole number. But that fact alone makes x pretty interesting! So we have reached a contradiction, from which we conclude that, actually, all whole numbers are interesting.

These are humorous examples, but paradoxes can sometimes be serious business. The most interesting paradoxes straddle the line between entertaining puzzles and deep philosophical questions. It did not seem proper to write a whole book about logic puzzles without devoting at least a few pages to this subject. So, let us consider a few of the more famous examples of paradoxes.

16.1 What Is a Paradox?

Philosopher R. M. Sainsbury writes:

> This is what I understand by a paradox: an apparently unacceptable conclusion derived by apparently acceptable reasoning from apparently acceptable premises. (Sainsbury 2009, 1)

Let us adopt this as our working definition.

It is important to distinguish paradoxes from mere contradictions. For example, you might be familiar with the so-called "Barber Paradox," which goes like this: Imagine that in a certain town, the barber shaves all those, and only those, who do not shave themselves. We now ask, "Who shaves the barber?"

If the barber does not shave himself, then he must be among those who are shaved by the barber, but since he is the barber, that means he *is* shaving himself. But if he does shave himself, then we have an instance of the barber shaving someone who shaves himself, which again contradicts the given information.

This is all very clever, but it is not a paradox. We have simply described a contradictory situation. As a matter of logic, it is not possible to have a town whose barber behaves in this manner.

A similar situation arises from the old teaser about what happens when an irresistible force meets an immovable object. By the former term, we mean a force so powerful that it sweeps aside all objects, while by the latter term, we mean an object that can remain motionless in the face of any force. No paradox arises from considering what happens when an irresistible force meets an immovable object. Instead, it is just contradictory to suppose that both could exist. If there is such a thing as an irresistible force, then that implies that there is no such thing as an immovable object, and vice versa. No paradox, just a contradiction.

We should also distinguish a paradox from a situation that is counterintuitive. An example is the Monty Hall problem, which many references refer to as the Monty Hall paradox. The problem is this: Imagine you are a contestant on a game show. You are shown three identical doors, one of which conceals a car, while the other two conceal goats. You select one of the doors, but you do not open it. The host of the show, whose name is Monty Hall, then opens a door, different from the one you chose, that he knows to conceal a goat. He now gives you the options of either sticking with your original choice, or of switching to the one other unopened door. You make your decision, and then win whatever is behind your final choice. What should you do, assuming you want to maximize your chances of winning the car?

The Monty Hall problem is often described as a paradox, because nearly everyone initially gives the wrong answer. Imagine that you choose door one, and Monty then opens door two, showing you a goat. Most people reason like this: "Now that door two is eliminated, there are two doors in play and each is equally likely to conceal the car. So it is a 50-50 choice, and it does not make a difference whether I stick or switch."

In reality, the correct reasoning is this: "If the car is behind door three, then Monty was forced to open door two, since he cannot reveal the car and he cannot open the door I chose. But if the car is actually behind my current choice of door one, then it was 50-50 whether he would open door two or door three. So Monty is twice as likely to open door two when the car is behind door three than he is when the car is behind door one. Therefore, I will double my chances of winning by switching doors."

For most people, serious thought is required to understand why the 50-50 solution is not correct. That notwithstanding, the Monty Hall problem is not a paradox. The principles of probability governing the scenario are perfectly well understood, and they are not mysterious to anyone with a proper education in the subject. The situation is not paradoxical, it is just difficult to grasp.

The sort of paradoxes that keep philosophers busy are different from this. They have in mind situations where the steps of the argument *really* seem unassailable, but the conclusion *really* seems unacceptable. When such a situation arises, it suggests there is some flaw in our understanding of the concepts underlying the paradox. It then falls to the philosophers to refine our understanding so that the paradox can be defused.

The philosophers have their concerns, while the rest of us have ours. The best paradoxes have undeniable entertainment value. For the audience to whom the paradox is presented, the effect is much like seeing a magic trick. You are following the action, thinking you understand perfectly what is going on, when suddenly something impossible happens. You laugh at how easily you were fooled. You then try to figure out how the trick was done and laugh all the harder when it resists your attempts.

16.2 Paradoxes of Predication

Paradoxes are especially intriguing when they arise from concepts that seem so simple and uncontroversial that you would think they could not possibly lead to trouble. An example is *predication*, by which we mean, essentially, the act of describing things by listing their properties.

(Note that in the philosophical literature, it is common to distinguish paradoxes of predication from paradoxes related to sets. They are, however, closely related, and I have found it convenient not to bother with that distinction.)

For any given predicate, it feels natural to discuss the set of things described by that predicate. For example, no conceptual difficulties arise from discussing the set of all red things, or the set of all people who live in the United States. In mathematics, we might discuss the set of even numbers, or the set of prime numbers, or perhaps the set of perfect squares between 100 and 1000 inclusive. Even if we give a contradictory description, such as the set of all numbers that are simultaneously even and odd, we do not thereby produce any conceptual confusion. A set whose defining condition is contradictory simply has no elements.

It is hard to imagine anything more fundamental than this. For any description you give, some things answer to it and some do not. What could go wrong?

Plenty, as it turns out. Consider this example from Bertrand Russell, who attributes its discovery to G. G. Berry:

> The number of syllables in the English names of finite integers tends to increase as the integers grow larger, and must gradually increase indefinitely, since only a finite number of names can be made with a given finite number of syllables. Hence the names of some integers must consist of at least nineteen syllables, and among these there must be a least. Hence 'the least integer not nameable in fewer than nineteen syllables' must denote a definite integer; in fact, it denotes 111,777. But 'the least integer not nameable in fewer than nineteen syllables' is itself a name consisting of eighteen syllables; hence the least integer not nameable in fewer than nineteen syllables can be named in eighteen syllables, which is a contradiction. (Russell 1908, 223)

Doesn't that just make you smile? How terribly clever to have noticed such a thing.

Russell himself is the author of the most famous paradox of predication. He noted that some sets answer to their own descriptions, while others do not. The set of all abstract ideas is itself an abstract idea, and therefore it belongs to itself. The set of all things that are not watermelons is not a watermelon, and the set of all sets with more than ten elements is itself a set with more than ten elements. In each case, the set answers to its own description. It is more common for sets not to answer to their own descriptions. The set of all red things is not itself red, and the set of all even numbers is not an even number.

In light of this observation, it ought to be coherent to define the set whose elements are precisely the sets that are not elements of themselves. Let us call this set R, in Russell's honor. We now ask whether R

is an element of itself, and you have probably already guessed what is to come:

- If we suppose that R is an element of itself, then that very supposition implies that R does not satisfy its own admission criterion. Recall that to be allowed into R, it was required that you not be an element of yourself.
- However, a similar problem arises if we suppose that R is not an element of itself. In this case, R actually *does* satisfy its own admission requirement and therefore is an element of itself. We get a contradiction either way.

This is serious business. The barber paradox described a contradictory situation, but we did not worry about it, since no one ever thought such a barber existed anyway. We do not have that option for dealing with Berry's or Russell's paradox. The predicates they consider seem perfectly reasonable. At every step we felt like we were playing by the rules, serving up coherent descriptions of actual objects, only to have the rug pulled out from under us at the last moment. To the extent that predication is a fundamental notion in logic and mathematics, it behooves us to have as clear an understanding of it as possible. These paradoxes show that our intuitive notions of these concepts are insufficient.

At this point, two general strategies are pursued. The first is to provide a proper set of axioms for set theory. We need to write down strict rules for what does, and does not, count as a proper predicate. This is not easily done. The reason we are studying sets at all is that we want to use them for specific purposes in mathematics. What is needed are rules that not only allow us to do what we want to do, but also rule out the dysfunctional predicates that lead to contradictions. This is tricky, since the line between what is legitimate and what is dysfunctional is often very thin. Nonetheless, it can be done, with the results presented in numerous unreadable textbooks on display in some of the dustier corners of the world's university science libraries.

The alternative approach is to praise Berry and Russell for their cleverness, and then to go on like before as if nothing had happened. This approach is sometimes called "naive set theory." Most mathematicians do not treat sets as objects of study in their own right, but rather use set-theoretic terminology and formalism to express other ideas. So long as you do not get too ingenious with your predicates, you will never run into trouble with this approach. This is the approach preferred by mathematicians in their professional lives.

My point is not that the axiomatics are unimportant. Instead, it is that they address concerns that are different from those that animate most working mathematicians. I noted in Chapter 8 that a major motivation for developing mathematical logic was the need to address perceived crises in the foundations of mathematics. This is true. But it is equally true that most mathematicians go their entire careers without ever worrying too much about foundations. Our confidence that mathematics is ultimately sound is not really based on any ruminations about the nature of formal systems. Instead, it comes from the simple observation that math works so well in practice. The centuries-long track record of practical successes convinces us that our methods ultimately rest on a solid foundation, even if it is difficult to give a perfect account of the nature of that foundation.

The history of calculus provides a good example of what I am talking about. When Isaac Newton first published his ideas about calculus, it quickly became apparent that there were many logical infelicities in what he had written. It is not an exaggeration to say that some of what he wrote flatly made no sense. Newton himself did not worry about this, however, and neither did the many scientists who subsequently applied his techniques to practical problems. They could see that calculus was producing good results, and this persuaded them that whatever logical difficulties there were in its theoretical formulations could be overcome. It fell to mathematicians in the eighteenth and nineteenth centuries to shore up the foundations. Their work proved very useful to those who subsequently sought to apply calculus in domains far removed from what Newton developed it for, but it is not as though everyone stopped using calculus while waiting for this foundational work to be complete.

At any rate, this is why we have the option of enjoying the paradoxes of predication just as amusing puzzles, rather than as anything that should send us into crisis and panic.

16.3   The Paradox of the Preface

Since predication seems to be a morass of unpleasantness, let us move on to a different fundamental concept: knowledge and rational belief. We have certain intuitions about what constitutes rational belief. For example, if you believe a proposition p, then you would generally be thought irrational if you also believed $\neg p$. Walt Whitman may have boasted, "Do I contradict myself? Very well, then I contradict myself. I am large, I contain multitudes," but this view is not generally thought conducive to sound reasoning.

Another principle that seems very reasonable is that if you rationally believe p and also rationally believe q, then it is rational for you to believe the conjunction $p \wedge q$. This is sometimes called the *principle of agglomeration*.

Incredibly, this seemingly self-evident principle can get you into trouble. Consider the following scenario, first presented by D. C. Makinson (1965): A scholar writes a lengthy book that presents a great many claims. She is a careful researcher who never makes a claim unless there is substantial evidence in its favor. At the same time, she is sufficiently humble to understand that it is effectively impossible to get everything right in such a massive undertaking. So, in the book's preface, she acknowledges that there is likely to be at least one false claim in the book. This suggests that she rationally believes each individual claim that she makes, but that she also rationally rejects the conjunction of all those individual claims.

We appear to have violated the principle of agglomeration. The scholar rationally believes each individual claim on the basis of her research, but also rationally rejects their conjunction on the basis of her past experiences of being wrong.

A classic paradox. Every link in the chain seems perfectly reasonable, but we are led to reject the principle of agglomeration, which also seems perfectly reasonable. What should we do?

Makinson's own solution was to make a distinction between a set of rational beliefs on one hand, and a rational set of beliefs on the other. To illustrate the difference, he defines some notation: Denote by $s_1, s_2, \ldots, s_n$ the various fact claims in the book, and let $b_1, b_2, \ldots, b_n$ represent the author's belief in each of these fact claims. Now define

$$s_{n+1} = \neg(s_1 \wedge s_2 \wedge \ldots \wedge s_n),$$

which is to say that s_{n+1} is the negation of the conjunction of the author's fact claims. Define b_{n+1} to represent the author's belief in s_{n+1}. Makinson now writes:

> When we say that it is rational to believe each of $s_1, \ldots, s_n, s_{n+1}$, we can be suggesting either of two things:
>
> (1) Each belief from the set $\{b_1, \ldots, b_n, b_{n+1}\}$ is rational.
> (2) The set $\{b_1, \ldots, b_n, b_{n+1}\}$ of beliefs is rational.
>
> In the case of the scholar and the preface, all that the example really yields is (1): each of the beliefs $b_1, \ldots, b_n$ expressed in the body of the book is rational, and so too is b_{n+1} expressed in the preface. The result (1) is perhaps rather surprising, but not paradoxical. We only enter into paradox when we read the ordinary English conclusion as (2), and mistakenly take a demonstration of (1) to be a demonstration of (2). (Makinson 1965, 207)

Makinson describes his solution as tentative, and that it surely is. The problem is that you can accept everything he is saying and still not believe that the paradox has been resolved. How does it come about that it is rational to hold each belief individually, but it is not rational to hold all of the beliefs simultaneously?

One way of making this idea more precise is to introduce probabilistic reasoning into the mix. This approach was first pursued by Lacey (1970), and was later refined by New (1978). The idea is this: When we speak of holding a belief based on reasons, we are rarely claiming that the reasons make the belief certain. Instead, we are claiming that the reasons justify feeling confident about the belief's correctness, while acknowledging there is still some possibility for error. That is, we assign a high probability to each belief individually. For simplicity, let us assume that the beliefs are independent of one another. In this case, standard probability theory says the probability of the conjunction of all of the beliefs is found by multiplying the individual probabilities together. But even if each individual probability is very close to 1, the product of all the probabilities could end up close to 0. In this way, we can make sense of why the individual beliefs are rationally held, but the conjunction of all of them is rationally denied.

Inevitably, many details need to be clarified in this account, but the strategy seems sound enough. New himself was certainly satisfied with this approach, concluding his paper with the blunt statement that the paradox had now been solved.

How adorable that he thought so. John N. Williams (1987) was not at all impressed with probabilistic approaches to this problem. He argued instead that we should bite the bullet and just reject the principle of agglomeration:

> I shall argue that counterintuitive appearances are not always deceptive; the preface case demonstrates the possibility of rational inconsistent belief. Attempts to deny the inconsistency of the case by giving it a probabilistic treatment misrepresent it. On the other hand, attempts to deny the rationality of the case by insisting upon the dependence of the available sets of evidence [the evidence for the beliefs individually and the evidence that the conjunction is false], stem from a crucial failure to recognise that the relevant beliefs are *inconsistent* rather than *contradictory*. (Williams 1987, 121)

Continuing the discussion would require getting into some very deep waters indeed, so we will pause it here. The Paradox of the Preface is just one of many so-called "epistemic paradoxes," which is to say, paradoxes that involve concepts related to knowledge. Just as in our discussion of predication, we find that the simplest assertions about knowledge or rational belief have the potential to plunge us into paradox.

Why can't anything ever be simple?

16.4 The Liar

Among all of the paradoxes that have been cataloged and studied, none has received as much attention as "the Liar." Some philosophers regard it as so intractable that they are willing to countenance the most eccentric revisions to our standard conceptions of logic just to address it. In various forms it has existed for as long as people have engaged in philosophical analysis.

The paradoxes of Berry and Russell show that naive ideas about predication lead to trouble. The Paradox of the Preface and others show that difficult questions surround our notions of knowledge and rational belief. At stake in the Liar paradox are fundamental intuitions about the nature of truth.

The simplest version of the paradox involves the following sentence, which I name L for convenience:

L: This sentence is false.

Let us refer to L as "the Liar sentence." We will soon encounter other sentences that, like L, do nothing more than make assertions about their own truth or meaningfulness; let us refer to them collectively as "Liar-type sentences."

The paradox arises when we try to determine the truth value of L. If we decide that L is true, then that means that what L asserts is actually the case. But since L asserts its own falseness, we would conclude that L is false. We have just shown that if L is true, then it is also false, which is a contradiction. Declaring L to be false fares no better. For if L is false, then what L asserts is actually seen to be the case, which would imply that L is also true. This is another contradiction.

Another mechanism for generating the same paradox is to consider the following two sentences:

(1) Sentence (2) is true.
(2) Sentence (1) is false.

I will leave it as an exercise to work out why these sentences lead to paradox.

At first blush, you might think there is a simple resolution: Perhaps L is neither true nor false. Many have tried this approach, but it runs afoul of the so-called "revenge liars." If you try to consign L to the badlands of the neither true nor false, then what will you do with the sentence: "This sentence is either false or neither true nor false"? Paradox arises regardless of whether you declare the sentence to be true, false, or neither true nor false. For that matter, how do you propose to handle: "This sentence is not true"?

We might now suspect that L is just defective in some way. For some reason, it is just not the sort of thing that receives a truth value. If we go this route, then it behooves us to spell out precisely what is defective about it. A very primitive attempt to do just that is simply to declare that self-reference is not allowed. It might seem that the problem with L is the fact that it refers to itself, prompting us to say: "New rule: no self-reference." The problem, though, is that many instances of self-reference seem entirely acceptable. For example: "This sentence is six words long." More seriously, we saw in Section 10.2 that self-reference lies at the heart of Gödel's first theorem, and we certainly do not wish to lose *that*. Thus, we are forced to come up with some nuanced theory of self-reference that preserves its good instances while discarding the bad, and good luck with that.

If simple solutions are not working, then perhaps we should consider more ingenious possibilities. Historically, an especially important approach to the Liar was suggested by Alfred Tarski. In his view, the Liar paradox revealed a genuine problem that besets English and indeed all natural languages. Specifically, our normal, everyday conception of truth is incoherent.

Let us spell out the problem: Normally, we think of true sentences as being the ones that describe the way things really are. When analyzing L, we said things like: "If L is true, then what it describes is actually the case." Using the sentence "snow is white" for the purposes of illustration, we can express this fundamental idea about truth by writing something like:

"Snow is white" is true if and only if snow is white.

The words "snow is white" appear twice in the above sentence. The first occurrence is in quotation marks, indicating that we are talking about the phrase itself, as an utterance. The second occurrence is not in quotation marks, indicating that we are using the same words to describe a state of affairs.

Tarski's point was that on one hand, this notion is fundamental to any normal conception of truth, but on the other, it leads to a contradiction when applied to L:

"This sentence is false" is true if and only if this sentence is false.

He argues that this incoherence shows that natural languages are inherently unsuitable for logical analysis and must be replaced by formal systems. In such systems, "truth" will be a predicate defined by whatever abstract rules we devise for it. The challenge is to define a predicate that captures our everyday notions of truth while avoiding the problems posed by Liar type sentences.

To carry out this project, he imagined that sentences are arranged in a hierarchy. At the lowest level are the normal sorts of sentences that people say to one another, which make no reference at all to semantic concepts like truth or meaning. Call this level zero. The next level up contains those sentences that predicate semantic concepts of those on the lowest level. For example, "snow is white" is a sentence at level zero, while "'snow is white' is true" is a sentence at level one. In general, the sentences at level n predicate semantic concepts of those at level $n-1$. The idea is that for a sentence to be meaningful, it can only predicate semantic concepts of sentences on lower levels. The liar sentence, L, violates this rule and is therefore meaningless.

As Tarski saw things, each level should be seen as employing a separate language from each of the levels below it. Even if in practice we use the same words (as we did in writing "'snow is white' is true if and only if snow is white"), we should still be clear that the second occurrence of "snow is white" is performing an entirely different linguistic function from the first occurrence. This leads to a distinction between the language in which our basic propositions are expressed, and the *metalanguage* we use to discuss the semantic properties of our basic propositions.

This is certainly one approach to the Liar paradox. A more recent, and more radical, approach goes in a different direction altogether. It suggests that we just bite the bullet and accept that L is simultaneously true and false. After all, we initially assumed it was true and derived a contradiction. This would generally be taken as an acceptable proof by contradiction that L is false. Then we assumed it was false and again derived a contradiction. This is a proof that L is true. So L is both true and false. Done!

Of course, this conclusion implies that the conjunction $L \land \neg L$ is actually true, which is to say that we have an example of a true contradiction. Most of us would probably balk at this point, arguing that if a theory forces us to accept true contradictions, then that theory is wrong and that is all there is to it. Philosophers of heartier stock demur, however, and they argue that true contradictions are perfectly acceptable. The view that there can be true contradictions is known as *dialetheism*; we mentioned it briefly in Section 1.2.

Even if we are willing to accept true contradictions as a possibility, we must not think of this as an explanation for free. We have a long road ahead of us if we want to make this idea work. Recall that in classical logic, a true contradiction implied any proposition at all, and the thought that every proposition is true really is a bridge too far. Thus, a dialetheist is committed to using a paraconsistent logic. Presumably, though, we do not want to discard classical logic altogether. Just look at all the good it has done! So some balance will have to be sought, and some philosophers have happily undertaken this task.

Many other solutions to the Liar paradox are on offer in the literature, most of them very elaborate and difficult. Tarski's hierarchy amounts to saying: "New rule: No talking about your own semantic properties," but this rule has little to recommend it beyond its ability to avoid paradoxes. Moreover, it suggests that people, who routinely engage in conversations about the truth and falsity of other people's utterances, are just systematically confused about what it is that they are doing.

As for dialetheism, did I mention that it involves true contradictions?

At this point, I must confess to being a bit befuddled by the ponderous scholarly literature on the Liar paradox. What, exactly, do the philosophers think they have accomplished with all this? Their relentless pearl-clutching over Liar-type sentences seems a bit overwrought, to put it kindly. Are we really forced to discard our most fundamental and intuitive notions about truth and noncontradiction because of: "This sentence is false"? Must we accept that we lowly peons need the Tarskis of the world to explain truth to us? Do we really have to consider pitching a tent on the blighted hellscape of dialetheism?

Must we endlessly waste time wrestling with every crazy idea philosophers dream up over their morning coffee?

That is hard to believe. It is even harder to believe, when you consider that there really does seem to be a very simple solution to the Liar paradox. Way back in Section 1.2, we discussed a distinction between sentences and propositions. It was propositions, not sentences, that are the bearers of truth. I have not mentioned that distinction since then, because most of the time it is unimportant. In normal discourse, the relationship between the sentences we say and the propositions they express is so close that we have no reason to emphasize the distinction. But if you are going to come at me with a weird, self-referential sentence, one whose meaning (much less its truth) is very murky, then I reserve the right to remind you that it is propositions, not sentences, that are truth bearers.

We hardly need Liar-type sentences to show us that there are grammatically correct, declarative sentences that do not express propositions. In linguistics, a famous sentence devised by Noam Chomsky says: "Colorless green ideas sleep furiously." The point is that grammatically, this sentence is acceptable, but it is nonsensical nevertheless. We might also point to sentences that are context dependent. "It is raining" is a perfectly good sentence, but until we stipulate a time and a place, it has no truth value. And some sentences lack a truth value because they fail to refer to anything. To illustrate this idea, Bertrand Russell used the example: "The present King of France is bald."

The solution to the Liar paradox is that sentence *L*, along with all the revenge liars, do not express propositions. This possibility has occasionally been floated in the literature. William Kneale expressed the point clearly:

> The moral to be drawn from the Liar paradox is not, as [Tarski] says, that ordinary language is radically inconsistent because it contains the words 'true' and 'false,' but rather that utterances must not be assumed to be specimens of propositions, and therefore either true or false, merely because they are specimens of grammatically well-formed sentences. (Kneale 1972, 241)

In a different paper, he was even more blunt: "[T]he paradox of the Liar holds no terrors for those who realize how the notion of truth is related to that of a proposition" (Kneale 1971, 321).

Yehoshua Bar-Hillel also emphasized this point (using the word "statement" in the way that we are using "proposition"): "Truth and falsity ... apply directly only to *statements* and not to (declarative) sentences. (Bar-Hillel 1957, 3). Later, he elaborates:

> It is indeed quite often perfectly safe to deal not with certain statements but with the sentences that are usually uttered when one wants to make these statements. In scientific matters especially it does little harm and a lot of good to work with sentences and assign them truth-values, so long as it is kept in mind that this is a matter of convenience only and that one must be prepared to return to statements as soon as trouble arises. But this procedure is suicidal when working with context-dependent sentences. ... [T]he proper conclusion from these failures is not that ordinary language is inconsistent, but that when the application of the rules of logic to ordinary language sentences leads to paradox, one has to fall back on statements. In the case of the Liar at any rate, the paradox disappears. (Bar-Hillel 1957, 5)

Prior (1958) and Garver (1970) also raise this possibility as a solution to the Liar.

If this solution works, then we should declare victory and go home. It turns out that we do not need to rewrite all the rules of logic and common sense, in the face of a few asinine, self-referential sentences. However, most philosophers reject this approach as a legitimate solution to the Liar. So far as I can tell, there is precisely one reason for this rejection. Susan Haack writes:

> This kind of approach suffers, I think, from inadequate explanatoriness— it doesn't supply a suitable rationale for denying the offending sentences a truth-value. Even granted for the sake of argument that only statements or propositions can be either true or false ... one would need an argument why in the case of the Liar one does not have an item of the appropriate kind.

After all, the Liar sentence suffers from no obvious deficiency of grammar and vocabulary. The minimum requirements would be, first, a clear account of the conditions under which an utterance of a sentence constitutes a statement; second, an argument why no utterance of the Liar could fulfill these conditions; third, an argument why only statements can be true or false. Otherwise one is entitled to complain that the solution is insufficiently explanatory. (Haack 1978, 140)

Haack protests that this approach—of distinguishing sentences from propositions or statements—is insufficiently explanatory. Interestingly, beyond a bald assertion that the Liar sentence suffers from no obvious deficiency of grammar and vocabulary, she gives no reason for thinking the approach is wrong. Since the alternative approaches all seem to entail some form of lunacy, perhaps we should not cavil about just how explanatory we are being.

Stephen Read agrees with Haack:

What we really need in order for this to be a convincing solution, is a proper account of meaningfulness relative to which it is clear that the sentence, 'This proposition is not true', is meaningless, and why. For that sentence certainly appears to be meaningful—we know what it means, namely, that the proposition it expresses is not true. (Read 1995, 153)

Read refers very casually to "the proposition it expresses," but whether or not it expresses any proposition at all is precisely the point at issue. If it does not express a proposition, then what we have is a simple failure to refer.

Matti Eklund was even more blunt:

I am impressed, however, with the following feature of the [L]iar paradox. Not only are we all taken in by each of the steps, but someone who failed to see that each of the steps in the reasoning is attractive would be agreed as having failed to grasp something about the concepts involved. (Eklund 2002, 322)

Eklund is presumptuous in saying that we are all taken in. I have never met a mathematician who finds the Liar paradox to be interesting or thought-provoking, and not for lack of looking. I would argue, contrary to Haack, Read, and Eklund, that Liar-type sentences do not at all look meaningful and do suffer from an obvious deficiency of vocabulary. At a minimum, we can say that words are being used in very unusual ways. When a sentence describes some state of affairs out there in the world, I feel like I understand what it means to say the sentence is true or false. But if a sentence does not describe such a state of affairs, but instead just refers, cryptically, to its own truth, then I no longer understand what it means to assign it a truth value.

If someone asked you to assign a truth value to "This sentence is green," you would just stare at him because the request does not make sense. You would at least ask for guidance about how to assess the color of a sentence before attempting to answer. Likewise for the sentence L. I cannot determine whether L is true until I first understand what L means. But I cannot understand what L means until I first understand the conditions that would make L true. (A similar point is made by Smullyan (1978, 217).) Does that not suggest that there is something defective about L?

Normally, we assess the truth of a sentence by first understanding the proposition it expresses and then comparing it to the aspect of reality to which the proposition relates. Understanding the language entails that you are capable of carrying out this assessment. In the case of L and all of the other Liar-type sentences, however, I have no idea how to describe the aspects of reality to which they relate, and I do not think anyone else does either. To anyone who regards these sentences as meaningful, please give me some clear account of their truth conditions. What would count as evidence one way or the other for their truth or falsity?

The Liar paradox arises only if you think truth is a predicate that must be applicable to all sentences of a given language. There is no reason for thinking that, however, and such a claim should not be part of anyone's intuitive understanding of how truth works. "Truth" and "falsity" are concepts invented by humans to describe the notions that some sentences describe the world as it is, while others do not. The former sentences are true; the latter are false. To the extent that we have any intuitions at all about truth, they are applicable only to the normal sorts of sentences that arise in daily conversation. No one has intuitions about the truth values of bizarre, self-referential sentences that arise only in philosophy seminars and make no intelligible claims about the world.

Haack and Read both stack the deck, egregiously, against the position I am outlining. They start from a presumption that Liar-type sentences are meaningful, so that when someone demurs, the burden is on them to provide a complete philosophical account for why they decline to get with the program. This is backward. The burden is on those who claim these sentences are meaningful to show that they assert something intelligible.

When we consider Liar-type sentences, we find first that they are not at all like the sentences to which we normally apply the concepts of truth and falsity. It is very murky what the truth conditions of such sentences could possibly be. Moreover, when we naively try to assign truth values to them, we are led to paradox. This certainly suggests that there is something defective about such sentences.

How should we respond? One approach is to suspect that these very weird sentences just express nothing at all, and then to go back to work as though nothing had happened. The alternative approach is to freak out, decide that all of our most fundamental and self-evident notions about logic are just completely mistaken, and then produce thousands of pages of unreadable analytic philosophy addressing the problem.

The former approach might be wrong, but it is the advocates of the latter who need to justify themselves.

16.5 Miscellaneous Paradoxes

Let me close this chapter by mentioning a few additional paradoxes that I think are clever. I will make no attempt to resolve the issues they raise, but instead will leave them as challenges for the reader.

In addition to the Russell and Berry paradoxes, another famous paradox of predication is attributed to Kurt Grelling. We notice that some words describe themselves and some do not. "Polysyllabic" is polysyllabic, and "noun" is a noun. Other words do not describe themselves, like "monosyllabic" or "verb." Let us say a word is "autological" if it describes itself, and "heterological" if it does not describe itself. We then ask if "heterological" is heterological. The analysis now proceeds in a manner similar to what we saw with Russell's paradox. If we decide that "heterological" is heterological, then it is a word that describes itself and is therefore autological. But if we decide instead that "heterological" is autological, then it is a word that does not describe itself, and is therefore heterological after all.

An especially well-known, and difficult, epistemic paradox is known as the unexpected examination. On Friday, a teacher tells her class that they will have an examination on one day of the following week. Moreover, on the chosen day, it will come as a complete surprise to them. Right up to the moment when they enter the classroom on the day of the examination, they will have no idea that it is coming. The students now reason that if Thursday comes and goes with no exam, then they will know that the exam is on Friday, since that is the only day left. This shows the exam cannot be on Friday. But if Wednesday then comes and goes with no exam, then they will know the exam is on Thursday, since Friday has already been eliminated as a possibility. Continuing in this way, they eliminate every day of the week and conclude that the teacher cannot possibly make good on her promise, and therefore they must not be having an exam at all. The following week they have the exam on Tuesday, and boy, were they surprised!

Finally, here is an ancient paradox. A famous lawyer takes on an apprentice and agrees to train him in the ways of the law. They strike a deal whereby the apprentice does not have to pay the fee for this service until he wins his first case. The apprentice then decides to avoid having to pay by never taking a case. Believing that this was not in the spirit of the agreement, the lawyer sues the apprentice to recover his fee. The lawyer reasons that if he wins the case, then he will recover his fee, since that is what the case is about. But if he loses the case, then his apprentice will have won his first case and will have to pay the fee anyway. The apprentice reasons that if he wins the case, then he does not have to pay, since that is exactly what the case is about. But if he loses then he has not yet won his first case and therefore does not have to pay anyway.

Take those as food for thought the next time you are on a long car ride or trapped in a dull committee meeting.

16.6 Notes and Further Reading

I opened Section 16.1 by quoting R. M. Sainsbury's definition of the word "paradox." The correctness of this definition has been challenged by Roy Sorensen (2003).

The books by Sainsbury and Sorensen to which I have alluded are both written at so high a level that they may be off-putting for a general audience. More lighthearted treatments of paradox can be found in the books by Smullyan (1980) and Clark (2012).

For a thorough discussion of the mathematical and philosophical issues surrounding the Monty Hall problem, see the book by Rosenhouse (2009).

My discussion of the Liar paradox treated it strictly as a paradox of natural language, and this is certainly in keeping with the way it is often discussed in the literature. However, it is also an issue when you are working in a formal system. In such a context, "truth" is just an abstract predicate that has nothing directly to do with correspondence with reality. The Liar paradox might then be telling us that predicates having certain desirable properties cannot be defined without lapsing into contradiction. As it happens, Raymond Smullyan was a great authority on self-reference in formal languages, and you can profitably consult his book on the matter to learn more (Smullyan 1994). The anthology edited by Martin (1970) is an interesting, if somewhat dated, collection of essays about the Liar paradox. More recent work in the area can be found in the anthology edited by Beall (2008).

The problem of the unexpected examination has an unusually large, and frequently difficult, literature associated with it. A good place to start is with the very readable discussion by Martin Gardner (1991). From there you can take a stab at the professional literature. Especially notable is the solution proposed by W.V.O. Quine (1953). Quine was one of the giants of twentieth-century logic, spending most of his career at Harvard, and his writing style usually reflected his awareness of his stature. His paper on this subject is called "On a So-Called Paradox," and it is written with the air of a weary father forced to deal with unruly children. The mere fact that he deigned to come down from Olympus to set everyone straight on this particular paradox, so-called or otherwise, suggests that it raises important issues. The general scholarly consensus regarding Quine's solution is that it works, but only for a relatively simple form of the paradox. It is possible to formulate more recalcitrant versions that resist Quine's solution. Of the many papers that have been written, three that I found interesting are those by Wright and Sudbury (1977), Chihara (1985), and Chow (1998).

CHAPTER 17

A Guide to Some Literary Logic Puzzles

We have encountered many great logicians to get to this point, but we have somehow omitted the greatest of them all. Of course, I am referring to Mr. Spock.

Spock, as he was usually called, was a fictional character on the 1960s science fiction television series *Star Trek*. He hailed from the planet Vulcan, whose residents are trained from a young age to suppress their emotions entirely. For them, only pure logic is important. Vulcans speak in a monotone, never smile or laugh, and calmly assess any situation without fear or anger clouding their judgment.

Spock was a capable officer, a brilliant scientist, and a loyal friend. On many occasions, he supplied the clever idea that saved his colleagues and defeated the bad guys. That notwithstanding, the show's human characters often treated him as deficient in some way. His logic came at the price of human feeling and compassion, you see, and this failing sometimes compromised his ability to lead the people under his command. For his part, Spock sometimes returned the hostility. In one episode he remarks: "May I say that I have not thoroughly enjoyed serving with humans? I find their illogic and foolish emotions a constant irritant."

In this final chapter, I would like to call your attention to certain works of fiction that, in my opinion, should be viewed as literary logic puzzles. The puzzles might be presented in the form of novels or short stories, but they are not the sort of thing you read for the memorable characters or penetrating social commentary. Instead, you read them for the sheer pleasure of seeing logic wielded as a weapon in the fight against evil. Growing up, I found many of these stories not just enjoyable but inspiring. I include this chapter in part to share with you some stories that have made an impression on me and that you might enjoy as well (having read to this point in a book about logic puzzles, after all).

There is a secondary point to be made, however. The stories I highlight here mostly hail from the genre known as the "classical detective story." Their protagonists are logic mavens whose main defining attribute is their ability to mentally record everything they see and to reason perfectly from this base of information. It is interesting that in nearly all cases, these characters are very much in the mold of Mr. Spock. The other characters might admire them for their abilities and devotion to justice, but there is often also an element of pity for their social awkwardness. It is the rare fictional character capable both of thinking clearly and of negotiating a social situation.

I have been careful not to include any major spoilers for the stories I discuss. These works will remain plenty enjoyable even if you go into them aware of the few tidbits that I mention. However, it has not been possible to reveal nothing at all, so proceed with caution.

17.1 The Nine Mile Walk

We need a term to describe the sort of fiction I will be spotlighting. If "science fiction" is fiction in which careful attention to scientific detail is paramount, then perhaps we could say that this chapter is dedicated to "logic fiction." That is, fiction in which impressive feats of logical reasoning are the point of the story.

In Section 1.2, I observed that working out the logical consequences of a set of premises can be surprisingly enjoyable. The most powerful illustration of this principle in fiction is a short story called "The Nine Mile Walk," by Harry Kemelman. It was first published in 1947.

The story is a dialogue between an unnamed narrator and an English professor named Nicky Welt. In the course of a casual conversation over breakfast at a local cafe, the following exchange takes place:

> "Give me any sentence of ten or twelve words," [Welt] said, "and I'll build you a logical chain of inferences that you never dreamed of when you framed the sentence."

> Other customers were coming in, and since the space in front of the cashier's booth was small, I decided to wait outside until Nicky completed his transaction. . . .

> When he joined me on the sidewalk I said, "A nine mile walk is no joke, especially in the rain."

> "No, I shouldn't think it would be," he agreed absently. Then he stopped in his stride and looked at me sharply. "What the devil are you talking about?"

"It's a sentence and it has eleven words," I insisted. (Kemelman 1967, 14–15)

From here Welt makes good on his promise and begins drawing inferences.

Difficulties arise quickly, however, because the sentence is presented with no context. This becomes an issue when Welt notes that the speaker did not say: "A nine mile walk in the rain is no joke." Instead, the sentence implies that a nine mile walk all by itself is no joke, and the rain only makes it worse. From this, Welt infers that since nine miles is not really *that* long a walk, the speaker is not athletic or an outdoors enthusiast. The narrator objects:

"Well, that would be all right under ordinary circumstances," I said, "but there are other possibilities. The speaker might be a soldier in the jungle, in which case nine miles would be a pretty good hike, rain or no rain."

"Yes," and Nicky was sarcastic, "and the speaker might be one-legged. For that matter, the speaker might be a graduate student writing a Ph.D. thesis on humor and starting by listing all the things that are not funny. See here, I'll have to make a couple of assumptions before I continue." (Kemelman 1967, 16)

The assumptions are that the sentence refers to a walk that was actually taken by someone whose motives were not frivolous (he was not just trying to win a bet, or anything like that) and that the walk was taken in the general area of the country where the story is taking place.

Given these assumptions, the chain of inferences leads to surprising places, ultimately building up to a *very* unexpected conclusion.

I sometimes assign this story early in the semester to the students in my math classes as a lighthearted way of introducing the importance of logical reasoning. The effects are sometimes polarizing. Half the class finds the story brilliant, while the other half finds it too implausible to be enjoyable. In my view, the former half has a promising future in mathematics, while the latter half is missing the point. You can certainly complain that some of Welt's inferences are far from airtight, and the whole thing *is* very implausible. The fun, though, is in seeing the way in which Welt builds one plausible inference on another, all starting from a sentence from which most people would infer nothing much at all. He engages in this academic exercise not because he knows going in that it will lead anywhere, but just for the intellectual satisfaction of taking a line of thought to its logical conclusion.

This is a good metaphor for mathematical work in general, with the important difference that in a chain of mathematical inferences, every link must be deductively certain. It can be very difficult to persuade students that, when working on a difficult problem, it is not necessary to have the whole flow of the solution in your mind before writing anything down. Instead, the point is to experiment and to play with the relevant concepts, just to see where the investigation leads.

17.2 The Early Days of "Logic Fiction"

Let us turn now to three pioneers in the field of logic fiction. The first is Voltaire, who, in a throwaway incident in one story, introduced several tropes that recur frequently in the genre. We then turn to Edgar Allan Poe and Arthur Conan Doyle, undoubtedly two of logic fiction's most famous practitioners.

In 1748, Voltaire published his novella *Zadig, or the Book of Fate*. The title character was able with a sword, but he was also pure of heart and bursting with good intentions. He was frequently naive regarding the treachery of those around him, however, and often found himself in trouble because of it. One such incident plays out like this:

> One day, as [Zadig] was out walking near a little wood, he saw running towards him the Queen's eunuch, followed by several officers who seemed in the grip of extreme anxiety and were running about distractedly as though trying to find some infinitely precious object. 'Young man,' said the chief eunuch, 'have you by any chance seen the Queen's dog?' Zadig modestly replied, 'It's a bitch, not a dog.' 'You are right,' said the chief eunuch. 'It is a tiny spaniel,' Zadig went on. 'It whelped recently, has a limp in the left front paw, and has very long ears.' 'You've seen it, then?' said the chief eunuch, completely out of breath. 'No,' Zadig replied, 'I've never seen it, and I didn't even know the Queen had a dog.' (Voltaire 1966, 8)

Shortly after this incident, one of the King's horses goes missing, and the master of the royal hunt asks Zadig whether he has seen the horse go by. Zadig replies:

> It is your best galloper. It is fifteen hands high, and has very dainty hooves. Its tail is three and a half feet long. The bosses on its bit are made of twenty-three carat gold, and it is shod with almost pure silver. (Voltaire 1966, 9)

The master of the hunt believes Zadig knows these things from having seen the horse, but Zadig professes that he has not.

Zadig is immediately arrested, since it is assumed that he knows such things because he has actually stolen the animals. He is sentenced to be whipped and imprisoned, but just after the sentence is passed, the lost animals are found unharmed. The sentence is commuted to 400 ounces of gold, on the grounds that while Zadig was not a thief, he plainly lied about not having seen the animals.

It is at this point that Zadig is allowed to speak for the first time. We learn that his inferences about the animals were based on his careful study of nature. When Zadig was approached by the eunuch, he noticed tracks in the sand that were plainly made by a small dog. Other marks in the sand indicated the long ears, as well as hanging udders, indicating the dog had recently given birth. One of the paw prints was not as deep as the others, indicating the dog was lame. The inferences regarding the horse were likewise made from a minute analysis of its tracks.

The incident now comes to a humorous conclusion:

> All the judges admired Zadig's profound and subtle discernment. Even the King and Queen came to hear it. Zadig was the one topic of conversation in the antechambers, in public and private, and although several magi thought he should be burned alive as a sorcerer, the King commanded him to be repaid the fine of four hundred ounces of gold he had been sentenced to pay. The clerk of the court, the ushers, the attorneys, called on him in full regalia to return his four hundred ounces. They just kept back three hundred and ninety-eight to cover legal costs, and their servants asked for payment too. Zadig realized how dangerous it could be to know too much, and made up his mind that next time he would tell nothing of what he saw. (Voltaire 1966, 10–11)

In the context of Voltaire's story, this is an isolated incident. Zadig does not again do anything like this in the remainder of the lengthy tale. However, the influence of this short scene on later writers was immense. Zadig's reasoning abilities are entertaining, and perhaps even inspiring for the reader, but they make Zadig himself a social outcast.

As influential as Voltaire was, it was not until much later that full-blown logic fiction was born. The first proper representative of this genre was Edgar Allan Poe's short story "The Murders in the Rue Morgue," first published in 1841. It is here that we have for the first time a story whose sole interest is in the process of reasoning through which the detective solves the puzzle. Poe's detective character is C. Auguste Dupin, though we are stretching definitions somewhat in describing him as a "character." We learn almost nothing about Dupin as a human being. He exists only as a sort of perfect reasoning machine.

The story is centered on a double homicide in a private residence in what was normally a quiet street in Paris. It is a crime of exceptional violence, with one body found thrust up a small chimney and the other found nearly decapitated on the grounds outside. Moreover, the murders appear to have taken place in an upstairs room that was locked on the inside. Witnesses heard the murderer speak inside the room, but could not agree on the language that was spoken. When the door was broken down, the murderer had vanished.

The police are helpless before the seemingly impossible situation. Dupin, by contrast, solves the puzzle just from reading the facts in the paper and carrying out a perfunctory investigation. He notes that it is precisely the exotic nature of the crime that makes it so easy to solve. Normalcy, he argues, has many explanations, but strangeness has only a few.

Among scholars of this genre, it is commonplace to note that Poe introduced numerous plot elements that quickly became clichés. The police are presented as competent in the grunt work of their job, collecting witness statements and whatnot, but they have no ability to reason creatively from what they find. Poe employed a first-person narrator to whom Dupin explains his reasoning, a character reminiscent of the more famous Dr. Watson from the Sherlock Holmes stories. That Poe introduced these tropes is true, but the story should also be remembered for its powerful opening statement:

> The mental features discoursed of as the analytical, are, in themselves, but little susceptible of analysis. We appreciate them only in their effects. We know of them, among other things, that they are always to their possessor, when inordinately possessed, a source of the liveliest enjoyment. As the strong man exults in his physical ability, delighting in such exercises as call his muscles into action, so glories the analyst in that moral activity which *disentangles*. He derives pleasure from even the most trivial occupations bringing his talent into play. He is fond of enigmas, of conundrums, or hieroglyphics; exhibiting in his solutions of each degree of *Acumen* which appears to the ordinary apprehension preternatural. (Poe 2006, 3)

This paragraph is among the clearest statements I have come across of what motivates mathematicians to do what they do.

Dupin appeared in two subsequent stories: "The Mystery of Marie Rogêt," published serially in late 1842 and early 1843, and "The Purloined Letter," published in 1844. It will add little to our discussion to present the fine points of these stories, save for a bizarre moment near the end of "The Purloined Letter."

The story is this: A man has an incriminating letter that he uses to commit blackmail. The police have strong reason to believe he is keeping the letter in his home, but the most careful, minute search of the premises fails to locate it. The problem is presented to Dupin, who quickly realizes what the police have missed. He is then able to recover the letter, a feat for which he is handsomely paid.

The moment to which I refer occurs while Dupin is discussing his reasoning in the case. The following exchange takes place, with the first-person narrator speaking first:

> "The Minister, I believe, has written learnedly on the Differential Calculus. He is a mathematician, and no poet."

> "You are mistaken; I know him well; he is both. As poet *and* mathematician, he would reason well; as mere mathematician, he could not have reasoned at all, and thus would have been at the mercy of the Prefect." (Poe 2006, 94)

In a story devoted to the pleasures of ratiocination, this attack on mathematicians is rather surprising, a point made by the narrator himself in the ensuing dialogue. Dupin, for his part, doubles down on his charge. Showing more emotion than he does at any other point in any of the three stories, he discourses for several hundred words on what shoddy thinkers mathematicians frequently prove to be.

Dupin's criticisms are ultimately expressed with too little cogency to be worth dissecting. I mention the incident only because a strange antipathy toward mathematics seems to recur periodically in logic fiction.

Poe certainly had an influence on other writers, most notably on Fyodor Dostoevsky, who was openly admiring of Poe. The character of Porfiry Petrovich, who appears as the detective in Dostoevsky's novel *Crime and Punishment*, was loosely modeled after Dupin (Frank and Magistrale 1997, 102). Another fan of Poe's stories was Charles Dickens, who frequently wove elements of crime fiction into his own work (Haining 1996). Throughout the mid-nineteenth century, various writers explored the possibilities of logic fiction (Cassiday 1983).

It was with Arthur Conan Doyle, however, and his creation of Sherlock Holmes in 1887, that the genre really exploded. Doyle especially emphasized the social isolation so often presented as the price for uncanny reasoning ability. Holmes is utterly bored by the inanities of normal life and requires constant mental stimulation in the form of puzzles to solve. When that stimulation is lacking, he turns to cocaine and heroin.

I suspect that most of you reading this are so familiar with Sherlock Holmes that I do not need to belabor the details of his stories. For our purposes, it is noteworthy that they are laced with moments meant to imply a conflict between logic and emotion. For example, in the story "The Greek Interpreter," Holmes remarks that his brother Mycroft is more talented than he in the ways of logical deduction. Watson suggests Sherlock is just being modest. Holmes replies:

> I cannot agree with those who rank modesty among the virtues. To the logician all things should be seen exactly as they are, and to underestimate

one's self is as much a departure from truth as to exaggerate one's own powers. (Doyle 1975a, 167)

It does not occur to Holmes that there is a difference between apprehending things accurately on the one hand, and always stating the truth bluntly on the other. Logicians, apparently, have no need to worry about the feelings of others.

We should mention that Holmes's archenemy was Professor Moriarty, who is twice described in the Doyle stories as a mathematician of singular brilliance. We learn that Moriarty is the author of two books: *The Dynamics of an Asteroid* and a treatise on the binomial theorem, whose title is not explicitly given.

Now, given my description of these works as "logic fiction," we ought to pause for a moment to decide what sort of logic we have in mind.

In the stories themselves, Holmes's methods are routinely described as deductive. The first Holmes novel, *A Study in Scarlet*, contains a chapter called "The Science of Deduction." That chapter contains passages like the following, in which Watson comments on a magazine article that was actually written by Holmes, though Watson was unaware of that fact when making his observations:

> It struck me as being a remarkable mixture of shrewdness and of absurdity. The reasoning was close and intense, but the deductions appeared to me to be far-fetched and exaggerated. The writer claimed by a momentary expression, a twitch of a muscle, or a glance of an eye, to fathom a man's innermost thoughts. Deceit, according to him, was an impossibility in the case of one trained to observation and analysis. His conclusions were as infallible as so many propositions of Euclid. So startling would his results appear to the uninitiated that, until they learned the processes by which he had arrived at them, they might well consider him as a necromancer. (Doyle 1975b, 18)

That last remark is reminiscent of people's reaction to Zadig's feats of ratiocination.

Among commentators on these stories, it is sometimes thought clever to point out, in light of Holmes's own emphasis on the certainty of his conclusions, that, technically, his reasoning is not really deductive at all (Pigliucci 2012). He actually uses a sophisticated form of induction. When he deduces a person's trade, or his recent activities, from various stains or marks on his clothes or skin, he is relying on correlations he has noted from practiced observations. Certain sorts of stains or marks are usually the result of certain sorts of activities. With all of Holmes's inferences, alternative possibilities are always logically possible but nonetheless unlikely. This is the hallmark of plausible inductive reasoning, as opposed to reasoning that is strictly deductive.

This point is well taken, but it would be churlish to make too big an issue of it. Purely deductive reasoning plays almost no role in daily life outside the exertions of professional mathematicians or philosophers. However, if we think of deductive reasoning as a sort of ideal to which all reasoning aspires, then we can fairly say that Holmes's reasoning comes far closer to that ideal than most reasoning does.

17.3 A Gallery of Eccentric Detectives

The success of the Sherlock Holmes stories ushered in the so-called "Golden Age" of detective fiction. Suddenly it seemed as though every crime was being solved by a socially awkward thinking machine who lorded his superior intellect over the bumblers in the local police force.

Notable in this regard was the character of Joseph Rouletabille, who appeared in seven novels, starting with 1907's *The Mystery of the Yellow Room*. He was the creation of French writer Gaston Leroux, who is better remembered for having written *The Phantom of the Opera* in 1910. The main crime in *The Mystery of the Yellow Room* bears a certain resemblance to "The Murders in the Rue Morgue," in that a woman is murdered inside a locked room. Witnesses heard the killer inside the room, but he had vanished by the time the door could be broken down. Leroux, however, introduced an innovation of his own, when later in the book, the killer escapes from several pursuers, even though every possible avenue of escape seems to have been anticipated and cut off. The inclusion of this second seemingly impossible situation added an element of horror to the proceedings, as it came to seem as though the killer really was in possession of supernatural gifts.

Pride of place in any discussion of Golden Age detective fiction must surely go to Agatha Christie's creation of Hercule Poirot, who would eventually appear in 33 novels and 51 short stories. Poirot was often mocking of those detectives, like Sherlock Holmes, who were endlessly looking for physical clues like cigarette butts. He preferred to emphasize the psychological aspects of the crimes he investigated. That notwithstanding, Poirot was always lecturing those around him to use their "little gray cells," and his solutions to the crimes were usually models of cogent (if not airtight) reasoning.

He also once provided a perfect illustration of the so-called hypothetico-deductive model of scientific reasoning, in his second outing, *Murder on the Links*, published in 1923. In this scene, Poirot has been away for two days from the scene of the novel's first murder. During this time, a second body was discovered. Colonel Hastings, who played the role of Dr. Watson in the early Poirot stories, has just informed Poirot that a second murder had occurred in his absence. Poirot replies:

"What is that you say? Another murder? Ah, then, I am all wrong. I have failed. Giraud [Poirot's rival] may mock himself at me—he will have reason!"

"You did not expect it, then?"

"I? Not the least in the world. It demolishes my theory—it ruins everything—ah—no!" He stopped dead, thumping himself on the chest. "It is impossible. I *cannot* be wrong! The facts, taken methodically and in their proper order, admit of only one explanation. I must be right! I *am* right!"

...

"Wait, my friend. I must be right, therefore this new murder is impossible unless—unless—oh, wait, I implore you. Say no word—" ...

"The victim is a man of middle age. His body was found in the locked shed near the scene of the crime and has been dead at least forty-eight hours. And it is most probable that he was stabbed in a similar manner to M. Renauld, though not necessarily in the back....

"Poirot," I cried, "you're pulling my leg. You've heard all about it already."
...

"I was right then? But I knew it. The little grey cells, my friend, the little grey cells! They told me. Thus, and in no other way, could there be a second death. (Christie 1983, 127–128)

Among the more famous detectives of the Golden Age are Lord Peter Wimsey, created by Dorothy Sayers, and Philo Vance, created by S. S. Van Dine (the pseudonym of Willard Wright). Less well known are detectives like Rogan Kincaid, created by Hake Talbot (the pseudonym of Henning Nelms). Talbot's novel *Rim of the Pit* involves multiple "impossible" situations, making it all the more satisfying when an entirely nonsupernatural explanation is provided.

A complete survey of such characters would require a book of its own, but it is worth noting two subgenres of the eccentric detective model.

The first involves detectives whose fame rested not on their feats of logical reasoning but rather on their thorough command of forensic science. Notable in this regard is Dr. John Thorndyke, introduced by R. Austin Freeman in his 1907 novel *The Red Thumb Mark*. Thorndyke would go on to appear in 22 novels and 40 short stories. Thorndyke is an exception to the social pariah model of the logic maven, since in this novel, the first-person narrator describes him as "the handsomest man he had ever met."

The main action of *The Red Thumb Mark* concerns a bloody finger-print that is found in a safe that has been burgled. The fingerprint is identified as belonging to a specific man, and on this evidence alone, the man is arrested. That leads to this dialogue between the first-person narrator and Thorndyke, with Thorndyke speaking first. You know you are reading logic fiction when people talk like *this*:

> "But there is no such thing as a single fact that affords evidence requiring no corroboration. As well might one expect to make a syllogism with a single premise."
>
> "I suppose they would hardly go so far as that," I said, laughing.
>
> "No," he admitted. "But the kind of syllogism that they do make is this— The crime was committed by the person who made this finger-print. But John Smith is the person who made this finger-print. Therefore the crime was committed by John Smith."
>
> "Well, that is a perfectly good syllogism, isn't it?" I asked.
>
> "Perfectly," he replied. "But you see, it begs the whole question, which is, 'Was the crime committed by the person who made this fingerprint?'" (Freeman 2010, 86)

Also notable is Dr. Lancelot Priestley, who appeared in an extraordinary 72 novels between 1925 and 1961. He was the creation of John Rhode (the pseudonym of Cecil Street). The Priestley novels were so similar to the Thorndyke novels that they do not require separate discussion here, except that in addition to being a doctor, Priestley was also a mathematician.

The second subgenre is that of religious detectives. The most famous in this regard is undoubtedly G. K. Chesterton's character Father Brown, a Roman Catholic priest who appeared in 53 short stories between 1910 and 1936. Evangelical Christianity was represented by the character Uncle Abner, who appeared in 22 short stories by Melville Davison Post between 1911 and 1928. Both characters frequently gained insight from their religious traditions, but the ultimate explanations for the crimes always involved earth-bound perpetrators acting from entirely human motives.

An especially notable story in the Uncle Abner corpus is "The Doomdorf Mystery." Doomdorf is found shot in his bed inside a locked room, by a gun that is found on its rack in the room. Another character quickly confesses to the crime on the grounds that he prayed for Doomdorf's death and then the latter died. A second character also confesses. She made a doll that represented Doomdorf and then stabbed it in the heart. Both characters note that Doomdorf was found in a locked room, suggesting that he was done in by the supernatural forces they had unleashed.

Abner is entirely unimpressed with these confessions and quickly provides a natural explanation for the facts. Readers who do not share Abner's worldview might find him a bit too religious, but "The Doomdorf Mystery" is a tale to make any rationalist cheer.

Though he falls outside the period of the Golden Age, under the category of religious detectives we should also mention Rabbi David Small, the creation of Harry Kemelman (whom we previously met as the author of "The Nine Mile Walk.") Rabbi Small solved 11 mysteries between 1964 and 1996. He appeared in one additional novel that was devoted to discussions of Jewish theology.

The Golden Age came to an end in the 1940s, as authors like Dashiell Hammett and Raymond Chandler started producing successful works in the "hard-boiled" style of detective fiction. The emphasis in these stories was on gritty realism, as opposed to logical fireworks. As a transitional form, you can consider Rex Stout's stories about Nero Wolfe, who appeared in 33 novels and 41 shorter works between 1934 and 1975. Wolfe was very much in the eccentric detective mode. He was morbidly obese and spent most of his time physically inert in his luxurious New York apartment, tending to his orchids. His Watson was Archie Goodwin, who gathered facts and provided the muscle when the situation called for it. Stout skillfully combined elements of Golden Age ratiocination with hard-boiled elements of sex and violence.

Nowadays, logic fiction has fallen on hard times. Browse the mystery shelves of the local bookstore, and most of the worthy examples will have been written many decades ago. Much of the best work is currently being produced by Japanese authors, of whom my favorite is Keigo Higashino. His stories about physics professor Manabu Yukawa, known to his friends as Detective Galileo, are very much in the classical mode. Higashino is very innovative in that he manages to create clever and difficult puzzles from settings with a very small number of characters. I highly recommend his Galileo novels *The Devotion of Suspect X* and *Salvation of a Saint*, as well as his non-series novel *Malice*.

17.4 The Anti-Logicians

So dominant was the Golden Age model in the early decades of the twentieth century that it was not surprising that a backlash took place against it. Some authors responded to the prevalence of brilliant, but inhuman, detectives by having *their* crimes solved by something other than cold, irrefutable logic.

The vanguard of the resistance was undoubtedly E. C. Bentley's novel *Trent's Last Case*, published in 1913. The story centers on a powerful, and hated, businessman who is found murdered in his mansion. The baffled police call in amateur detective Philip Trent to aid in the investigation. When we first meet him, Trent seems to be very much in the Sherlock Holmes mold. His incredible attention to detail and prodigious talent for logical thinking have brought him success in several previous investigations. Soon enough, Trent propounds an ingenious solution to the crime, all based on seemingly unimportant observations regarding the disposition of the victim's shoes. Trent's solution is later shown to be wrong in several key respects, including in its identification of the guilty party. Eventually the actual criminal explains to Trent what really happened and where his reasoning had gone wrong, prompting Trent to announce his retirement from crime detection.

Also representative of the anti-logician school is Freeman Wills Crofts, who published his first novel in 1920. His series character was Inspector Joseph French of Scotland Yard. Unlike Sherlock Holmes or Hercule Poirot, French was a dull plodder who solved his crimes through sheer hard work. He rarely showed moments of impressive insight, and he was not keen on taking deductive leaps from seemingly insignificant clues. As representative, we can take Crofts's 1933 novel *The Hog's Back Mystery*. Three seemingly harmless people disappear under suspicious circumstances from a quiet British town and are eventually found murdered. Inspector French investigates, but he finds himself frequently stuck and frustrated. He eventually unravels the crime and brings the guilty parties to justice. This is accomplished not by intellectual derring-do but by a meticulous accounting of the movements of all the characters, coupled with detailed time tables of events and travel times between relevant locales.

Undoubtedly, though, the greatest proponent of this school was Anthony Berkeley. His series detective, Roger Sheringham, is like Holmes and the others in that he is an amateur with a gift for logical reasoning, but is unlike them in that his proposed solutions are frequently wrong.

Berkeley's most famous work was his 1929 novel *The Poisoned Chocolates Case*. As the novel opens, we learn of a prominent socialite who has been killed by being tricked into eating poisoned chocolates. The six members of an amateur detective club, including Sheringham, decide they will each investigate on their own and then report back with their solutions. In the end, each member presents a different solution, each more plausible than the one before, but all of them wrong until the final solution is offered.

Along the way, several characters make fools of themselves by an over-reliance on dubious inferences. One clue is the basis for four entirely distinct inferences by different members of the club. One club member outlines a very tenuous chain of inferences that leads her to accuse another member of being the murderer. Still another outlines a list of characteristics that the guilty person must possess. So detailed is the list, and so implausible the chance of finding all of the characteristics in one person, that he argues that finding the murderer is a matter of finding one person who possesses all of them. He then discovers that he possesses all of the characteristics. Rather than abandon his theory, he concludes he must actually be guilty, but has no memory of the crime due to some mental illness.

Suffice it to say that the real guilty party is eventually revealed, but that deductive reasoning as a method for solving crimes rather takes it on the nose.

17.5 Carr and Queen

Let us now discuss the work of the two most important authors of logic fiction: John Dickson Carr and Ellery Queen. First, though, we should consider some criticisms of Golden Age stories.

In 1941, Raymond Chandler published an essay in the *Atlantic* magazine. It opened like this: "Fiction in any form has always intended to be realistic" (Chandler 1950, 1). With an opening like *that*, you just know it is going to be a long night for the classical detective story.

Chandler was the creator of Philip Marlowe and a prominent representative of the hard-boiled school of detective fiction. He was mostly unimpressed with logic fiction and found many faults in its leading practitioners. He writes:

> It is the ladies and gentlemen of . . . the Golden Age of detective fiction that really get me down. This age is not remote. For Mr. Haycraft's [a literary critic of the time] purpose it starts after the First World War and lasts up to about 1930. For all practical purposes it is still here. Two thirds or three quarters of all the detective stories published still adhere to the formula the giants of this era created, perfected, polished, and sold to the world as problems in logic and deduction. (Chandler 1950, 5)

He proceeds to excoriate Golden Age writers for the utter implausibility of their plots and for how unoriginal the whole genre had become. He eventually builds up to this:

> There is a very simple statement to be made about all these stories: they do not really come off intellectually as problems, and they do not come off artistically as fiction. They are too contrived, and too little aware of what goes on in the world. . . . But if the writers of this fiction wrote about the kind of murders that happen, they would also have to write about the authentic flavor of life as it is lived. And since they cannot do that, they pretend that what they do is what should be done. (Chandler 1950, 11)

I have an image in my head of Chandler being dismissive of knight/ knave puzzles on the grounds that no one really talks like that.

Chandler was being a curmudgeon, but perhaps he has a point. I do not recommend reading Golden Age fiction for its literary value. No actual murderer would plot a crime as baroque as what these authors devised, and if they ever did, the very complexity of their plans would quickly lead to their downfall. The characters were seldom developed beyond the necessary minimum, and unsurprisingly, given the time in which they were written, frequently portrayed racial and sexual stereotypes that are jarring to a modern reader. Chandler is right that a talent for building complex puzzles is not often accompanied by a talent for sharp dialogue and proper pacing.

The fact remains, however, that Chandler is missing something fundamental, and it was expressed perfectly by John Dickson Carr, himself an eminent practitioner of Golden Age fiction:

> Now it seems reasonable to point out that the word 'improbable' is the very last which should ever be used to curse detective fiction in any case. A great part of our liking for detective fiction is *based* on a liking for improbability. When A is murdered, and B and C are under strong suspicion, it is improbable that the innocent-looking D can be guilty. But he is. If G has a perfect alibi, sworn to at every point by every other letter of the alphabet, it is improbable that G can have committed the crime. But he has. When the detective picks up a fleck of coal dust at the seashore, it is improbable that such an insignificant thing can have any importance. But it will. In short, you come to a point where the word 'improbable' grows meaningless as a jeer. (Carr 1935, 161)

You might have noticed from my brief survey that many Golden Age plots involved seemingly impossible situations: murderers escaping from locked rooms, letters hidden so ingeniously that the most minute search fails to reveal them, and so on. It is unsurprising that devotees of logic fiction would dwell so frequently on such situations, since they present the reader with a clear logical problem akin to a paradox. The basic facts of the case seem clear and well attested, but they lead to an impossible conclusion. The challenge to the reader is immediate: Which premise is wrong?

Carr wrote close to 70 novels in his life, and most of them involved impossible situations. Fifty-five of those novels featured one of his two series characters: Dr. Gideon Fell and Sir Henry Merrivale. His masterpiece is generally considered to be the Dr. Fell novel *The Three Coffins* (also published under the title *The Hollow Man*), which involves the remarkable murder of Professor Grimaud. Witnesses see the murderer enter Grimaud's study, and the door is subsequently under constant surveillance. A struggle is heard from within, and then a gunshot. When the room is entered, the murderer has vanished. There is a window in the room, but the snow on the ground below and on the roof above shows no footprints. Later, a second murder occurs at close range in the middle of a street, but only the victim's footprints mar the snow. How were the murders accomplished?

Like Poe and Doyle before him, Carr had a distaste for mathematics. His characters frequently disparaged the subject, such as in this quotation from his novel *Dark of the Moon*:

> To me mathematics means the activities of those mischievous lunatics A, B, and C. In my time they were always starting two trains at high speed from distant points to see where the trains would collide somewhere between. . . . When the silly dopes weren't wrecking trains or computing the ages of their children without seeming to know how old the brats were, two of 'em had a passion for pumping water out of a tank while the third poor mug pumped water into it. (Carr 1967, 61–62)

However, the author whose work is most aptly described as "literary logic puzzles" is surely Ellery Queen. It would be more accurate to say "authors," since Ellery Queen was the pseudonym for the team Frederic Dannay and Manfred Lee. Ellery Queen is also the name of their series detective character, and in the novels, he is a writer of detective stories. Ellery Queen the character is described like this:

> He was the pure logician, with a generous dash of dreamer and artist thrown in—a lethal combination to those felons who were so unfortunate as to be dissected by the keen instruments of his mind, always under those questing pince-nez eyeglasses. (Queen 1930, xii)

Their novels are truly logic puzzles in novel form, with Ellery frequently presenting his summations in the form of rigorous deductive arguments. Several of their early novels, easily recognized because they had titles like *The Roman Hat Mystery*, *The French Powder Mystery*, and *The Dutch Shoe Mystery*, were actually subtitled "A Problem in Deduction." Among many solid novels, Queen's standouts are *The Greek Coffin Mystery* and *Cat of Many Tails*.

17.6 The Thinking Machine

Which brings me, finally, to the Thinking Machine.

We have seen a great many amateur detectives in this chapter. These folks were basically superheroes, with their superpower being their freakish ability to observe everything and to reason perfectly. However, we have not yet encountered an actual logician.

The Thinking Machine was the nickname of Professor S.F.X. Van Dusen, a character created by Jacques Futrelle in 1906. He would eventually appear in 51 short stories and one novella. Like most of the sagacious reasoners we have encountered, The Thinking Machine was unusual looking and socially awkward. He is described as being perpetually annoyed with the people around him. He was prone to saying things like this:

> Two and two make four, not sometimes, but all the time. As the figure 'two' wholly disconnected from any other, gives small indication of a result, so is an isolated fact of little consequence. Yet that fact added to another, and the resulting fact added to a third, and so on, will give a final result. That result, if every fact is considered, *must* be correct. Thus any problem may be solved by logic; logic is inevitable. (Quoted in Ellison 2003, xxiii)

Among The Thinking Machine stories, there is one that is so much more compelling than the others that I have chosen to present it as the final item in this book. It is called "The Problem of Cell 13," and it is a strong candidate for the finest detective story ever written. My father showed it to me when I was in middle school, and it made a *big* impression.

The story opens with The Thinking Machine engaged in casual conversation with some friends. He puts forth the notion that there is no problem the mind cannot solve. His friends are unimpressed. They suggest that this is just idle talk, and that the mind, for example, could not think prison walls out of existence. The Thinking Machine notes that a person could so apply his intellect as to escape from a cell. This leads to the following exchange (the first speaker is Dr. Ransome, The Thinking Machine's friend):

> "Take a cell where prisoners under sentence of death are confined—men who are desperate and, maddened by fear, would take any chance to escape—suppose you were locked in such a cell. Could you escape?"

> "Certainly," declared The Thinking Machine.... "You might treat me precisely as you treated prisoners under sentence of death, and I would leave the cell."

> "Not unless you entered it with tools prepared to get out," said Dr. Ransome.

> The Thinking Machine was visibly annoyed, and his blue eyes snapped.

> "Lock me in any cell in any prison anywhere at any time, wearing only what is necessary, and I'll escape in a week," he declared, sharply. (Futrelle 2003, 17)

They are able to arrange for the experiment to begin that very night, with only the prison warden knowing that The Thinking Machine was not really a prisoner. Of course, the warden regards the whole thing as a joke and believes The Thinking Machine has no chance of making good on his boast. The Thinking Machine is then given a week to make his escape. Prior to starting the experiment, he instructs his housekeeper, in the presence of everyone involved, to have dinner ready for everyone exactly one week to the minute of his being locked in the cell.

Now, under the circumstances I do not think it counts as a spoiler to tell you that he does, in fact, make his escape. The method by which he does so is highly implausible, of course, but it is also ingenious and not in contradiction of any known principles of physics. To tell you more would constitute too great a spoiler, but I do wish to close with one more brief excerpt, which takes place right after The Thinking Machine makes his escape:

"How did you do it?" demanded the warden.

"You gentlemen have an engagement to supper with me at half-past nine o'clock," said The Thinking Machine. "Come on, or we shall be late."

"But how did you do it?" insisted the warden.

"Don't ever think you can hold any man who can use his brain," said The Thinking Machine. (Futrelle 2003, 39)

For a nerdy, introverted kid more interested in chess and mathematics than in sports, this was a powerful cri de coeur.

Friends, it is time to stop apologizing for liking this sort of thing. Pay no attention to the snobs and Pecksniffs and killjoys and Chandlers who will lecture you on the true nature of great literature. Let them hold forth on the importance of grim realism in works of fiction, as though the point of reading is to remind you of all the daily unpleasantness from which you have turned to books to escape. Let them strain their necks looking down on you, for it is they who are missing out.

It is hard to imagine any situation more hopeless than the one in which The Thinking Machine finds himself at the start of this story. Seriously, how on earth is he going to get out of that cell? Who would willingly put himself in such a position? Who would be so confident in his ability to reason, and in the ability of reason to surmount any obstacle? *That* is inspiring. *That is* great literature.

And it is the perfect place to wrap this up and say good night, for we have returned to where we started. The lesson learned from this brief survey of literary logic puzzles is precisely the one I have emphasized from the start. Reasoning is fun! There are few things in life as satisfying as encountering opacity and, by applying nothing more formidable than the power of your mind, replacing it with clarity. It is easy to forget that simple truth as you suffer through the tedium of a math class, or as you try to slog through the molasses-like prose of the average textbook.

But it is no less true for *that*.

GLOSSARY

agglomeration, principle of The principle of rational belief that holds that if a person rationally believes two propositions individually, then it is also rational for that person to believe the conjunction of those two propositions. See Section 16.3.

antecedent See *conditional statement.*

argument A sequential set of statements arranged so that the final statement is claimed to follow logically from the others. The final statement is called the *conclusion* of the argument, while the other statements are referred to as *premises.*

Aristotelian logic A branch of logic devoted to studying syllogisms, which are arguments with two premises and a conclusion. Pioneered by Aristotle in his work *Prior Analytics,* in roughly the 300s BCE. Aristotelian logic is also sometimes called "term logic."

axiom A statement accepted as true for the purpose of inaugurating some process of logical reasoning. The axioms of a formal system are analogous to the rules of a game such as chess.

Berry's paradox The paradox that arises from the description "the least integer not nameable in fewer than nineteen syllables." See Section 16.2.

biconditional statement A statement employing the "if and only if" connective.

biliteral diagram A sort of Venn diagram used by Lewis Carroll to represent the overlaps between two sets. See Section 4.2.

bivalence, principle of The logical principle that holds that every proposition has exactly one of the two truth values "true" or "false."

Caliban's Will An especially famous logic puzzle, first presented by L. R. Ford in 1947. See Section 15.2.

Carroll's regress A term referring to a puzzle presented by Lewis Carroll, in which he asks: What compels us to accept the logical conclusions of a set of propositions? If the compulsion is said to arise from our acceptance of some other proposition, we quickly arrive at an unacceptable infinite regress. See Sections 6.2, 6.3, and 6.4.

categorical proposition A statement of one of the following forms: All As are B, No As are B, Some As are B, Some As are not B. In Aristotelian logic, these are referred to as statements of type *A, E, I,* and *O,* respectively. Such statements are also referred to as *universal affirmatives, universal negatives, particular affirmatives,* and *particular negatives,* respectively.

categorical syllogism A syllogism in which both premises and the conclusion are categorical propositions.

classical logic The system of logic typically used in modern mathematics. Among other principles, it holds that every proposition is either true or false, but not both simultaneously.

coercive logic A humorous term coined by Raymond Smullyan for describing puzzles in which questions are asked or statements made that logically compel certain behaviors from those who hear them. See Section 11.6.

coherence theory A view of truth that holds, roughly, that a statement is true or false depending on whether or not it coheres with some other given set of statements. Contrast with *correspondence theory* and *redundancy theory*.

complement (of a term) The set of all things not in the *extension* of a given term. For example, the complement of the term "cats" is the set of all things that are not cats.

conclusion The final line of an argument. If the argument is valid, then the conclusion must be true if all the premises are true.

conditional statement A statement employing the connective "if–then." The clause following "if" is said to be the antecedent, while the clause following "then" is said to be the consequent. Conditional statements are sometimes referred to as hypothetical statements, or implications.

conjunction A statement employing the "and" connective. In logic, the statement "P and Q" is true when P and Q are both true individually, and it is false otherwise.

consequent See *conditional statement*.

contradictories Two propositions that must have different truth values. See Figure 3.2.

contraposition The operation that reverses and negates the terms of a categorical proposition. For example, the contrapositive of "All As are B" is "All non-Bs are non-As."

contraries Two propositions that cannot both be true, but which can both be false. See Figure 3.2.

converse See *conversion*.

conversion The operation that reverses the terms of a categorical proposition. For example, the converse of "All As are B," is "All Bs are A."

copula The form of the verb "to be" that connects the subject and predicate of a categorical proposition. For example, in "All cats are mammals," the copula is "are."

correspondence theory A view of truth that holds that true statements are the ones that correspond to the facts. This view is intuitively natural, but it is difficult to spell out precisely the nature of the correspondence relation, or even what is meant by "facts." Contrast with *coherence theory* and *redundancy theory*.

deductive reasoning Drawing a conclusion that follows as a matter of logic from given information. When reasoning deductively, we claim that the conclusion simply must be true if the given information is correct. Contrast with *inductive reasoning*.

deontic logic A system of logic in which concepts like "ought to" and "permitted to" are expressible.

dialetheism A view of logic that holds that there are true contradictions. See Sections 1.2 and 16.4.

disjunction A statement employing the "or" connective. In logic, the statement "P or Q" is evaluated to be false when P and Q are both false and true in all other cases. This does not capture every use of "or" in spoken English, but it is a useful convention for many purposes.

Eliminands A term used by Lewis Carroll to refer to the terms from a set of categorical premises that do not appear in the conclusion drawn from those premises.

Entity A term used by Lewis Carroll to refer to categorical propositions that assert the existence of something.

epistemic obligations A term used by philosophers to refer to the idea that in certain situations, we might be thought to be compelled to think in a certain way, on pain of lapsing into error or incoherence.

excluded middle, law of In classical logic, the principle that holds that for any proposition, either that proposition or its negation is true.

existential import A property possessed by a statement that implies the existence of at least one object. In modern treatments of Aristotelian logic, statements of the form "Some As are B" and "Some As are not B" are held to have existential import, while statements of the form "All As are B" and "No As are B" are held to lack existential import. (For example, the statement "All unicorns are horses" should not be taken to imply that unicorns exist.)

explosion, principle of The logical principle asserting that any proposition follows from contradictory premises. Classical logic is explosive in this sense. Contrast with *paraconsistent logic.*

extension (of a term) The collection of objects to which the term accurately applies. For example, the extension of the term "cat" would consist of all the specific animals in the world we classify under that term. Contrast with *intension (of a term).*

extreme terms In a categorical syllogism, the two terms that each appear in exactly one of the premises.

figure (of a syllogism) In a categorical syllogism, the arrangement of the middle terms with respect to the extreme terms. See Section 3.4.

formal logic A branch of philosophical inquiry based on the principle that arguments expressed in a natural language are made valid by their adherence to certain abstract argument forms. For example, the argument "All cats are mammals, all mammals are animals, therefore all cats are animals" is valid, because it is an instantiation of the valid form "All As are B, all Bs are C, therefore all As are C."

formal system A set of abstract symbols equipped with syntactical rules for assembling them into formulas, and with rules of inference for deriving new formulas from those previously established. See Chapter 9.

fuzzy logic A system of logic in which a continuum of truth values is allowed.

Game of Logic, The A book first published by Lewis Carroll in 1886. It essentially created the field of recreational logic. Carroll's puzzles in the book were based on Aristotelian logic.

Gödel's incompleteness theorems Two especially important theorems in mathematical logic proved by Kurt Gödel in the 1930s. The first theorem says that every formal system capable of expressing arithmetic must be capable of expressing statements that are true, but unprovable. The second says that the consistency of a formal system cannot be proved within the system. See Chapter 10.

Grelling's paradox The paradox that arises from defining "heterological" to describe words that do not describe themselves. A contradiction arises regardless of how we answer the question: "Is 'heterological' heterological?" See Section 16.5.

Hardest Logic Puzzle Ever, the A specific puzzle first discussed in detail by George Boolos in 1996. The puzzle involves three omnipotent gods, one who always tells the truth, one who always lies, and one who decides randomly whether to answer truthfully or falsely. The gods respond to questions in their own language, whose words for "yes" and "no" are "da" and "ja," but it is unknown which means which. You must determine who is who by asking three yes/no questions. See Chapter 14.

hypothetical statement See *conditional statement*.

immediate inference An inference drawn from a single premise.

implication See *conditional statement*.

indexical statement A statement whose meaning depends on the context. For example, the meaning of "I have a cat" changes, depending on the speaker.

inductive reasoning Drawing a general conclusion from a large number of specific instances. When reasoning inductively, we do not claim our conclusions are logically certain, but only that they are likely, given the data. Contrast with *deductive reasoning*.

instrumentalism In logic, the philosophical position that it does not make sense to assess systems of logic as correct or incorrect. Instead, they should be assessed as useful or not useful for some purpose.

intension (of a term) The set of properties or qualities connoted by the use of a term. For example, the intension of "cat" would include properties like "furry," "four-legged," and "mammal," among others. Contrast with *extension (of a term)*.

knights/knaves Characters in a certain type of logic puzzle. Knights only make true statements, while knaves only make false statements. Contrast with *normals*.

Liar paradox The paradox arising from trying to assign a truth value to the sentence: "This sentence is false." See Section 16.4.

logic The study of what inferences follow from given information.

logical constants A word, phrase, or symbol whose purpose is to establish logical relationships among the component clauses of a complex sentence. They are "constant" in the sense that they have the same meaning everywhere they are used. Examples of logical constants in English are: *and, or, not*, and *if–then*.

logical pluralism The philosophical position that there is no one universally correct system of logic.

logic fiction A term coined to describe literature whose main source of interest lies in the brilliant feats of logical deduction exercised by the protagonist. See Chapter 17.

major/minor premise In a categorical syllogism, the premise containing the predicate of the conclusion is the major premise, and the premise containing the subject of the conclusion is the minor premise.

major/minor term Respectively, the predicate and subject of the conclusion of a categorical syllogism.

many-valued logic Any system of logic in which there are more than two truth values. For example, a system of three-valued logic might allow statements to be assessed as either true, false, or neither true nor false. Contrast with *classical logic*.

mathematical logic A branch of mathematics that takes formal systems as objects of study in their own right.

metaphysics The branch of philosophy that studies the nature of being, and the nature of the sorts of worlds in which being can manifest itself. It is difficult to define precisely, but you tend to know it when you see it.

metapuzzle A term coined by Raymond Smullyan to describe a sort of logic puzzle whose solution depends on knowing that certain other puzzles either can or cannot be solved. See Chapter 15.

middle term In a categorical syllogism, the term that appears in both premises.

modal logic Any system of logic that includes the notion that there is a distinction between statements that are necessarily true versus those that are possibly true.

modus ponens The principle of inference asserting that from "p" and "if p, then q," you can validly infer q.

modus tollens The principle of inference asserting that from "if p, then q" and "not q," you can validly infer "not p."

mood (of a syllogism) A triplet such as *AAA* or *EAE*, representing the three types of statements used in a given categorical syllogism. See Section 3.4.

muddy children, the A classic puzzle in epistemic logic. See Puzzle 6.

natural language A language used in normal conversation. Examples include English, French, and German. Contrast with *symbolic language*.

negation (of a proposition) A proposition that asserts, of a given proposition p, that p is false.

Nelson Goodman principle The principle that on an island of knights and knaves, the truth of any proposition p can be determined by asking a native: "Is p true if and only if you are a knight?" Raymond Smullyan attributes the discovery of this principle to Nelson Goodman. See Section 11.4.

nominalism In philosophy, nominalism with respect to some class of objects is the view that the terms we use to describe those objects do not refer to anything real, but are just useful as names for some concept. In logic, it is generally the view that abstract objects do not actually exist, but that the words that refer to them are useful fictions. Contrast with *realism*.

nonclassical logic Loosely, any system of logic that rejects one of the foundational principles of classical logic, such as the principal of bivalence, the law of the excluded middle, or the law of noncontradiction.

noncontradiction, law of A principle of classical logic holding that a proposition and its negation cannot both be true simultaneously.

normals Characters in a certain type of logic puzzle. Normals sometimes make true statements and sometimes make false statements. Contrast with *knights/knaves*.

normative force In logic, the idea that the principles of logic impose certain requirements on our reasoning, on pain of being thought irrational or naughty if they are violated.

Nullity A term used by Lewis Carroll to refer to categorical propositions that assert the nonexistence of something.

obversion The operation that changes the quality and negates the predicate of a categorical proposition. For example, the obverse of "All As are B" is "No As are non-B."

ontology The branch of philosophy that studies the nature of being and tries to pin down what actually exists.

Organon, The Aristotle's six works on logic, treated as a whole.

paraconsistent logic Any system of logic in which it is not the case that any proposition follows from contradictory premises. Contrast with *explosion, principle of*.

paralogism An error made in the application of logic. Someone who claimed that "If p, then q" is logically equivalent to "If q, then p" would be guilty of a paralogism.

particular affirmative/negative See *categorical proposition*.

Peano arithmetic An axiomatization of standard arithmetic developed by Italian mathematician Giuseppe Peano in the late nineteenth century. See Section 9.3.

Port-Royal Logic, The A logic textbook first published in 1662 by Antoine Arnauld and Pierre Nicole. It is generally thought to have revived scholarly interest in logic after a long period of decline.

Preface, Paradox of the The paradox that arises from the fact that an author might rationally believe every factual claim in her book, while also recognizing that, because she is fallible, it is unlikely that the conjunction of all those factual claims is true. This seems to contradict intuitive principles of rational belief. See Section 16.3.

premise A statement providing information that will be used, together with other such statements, to draw a conclusion. See also *argument*.

Prior Analytics See *Aristotelian logic*.

proposition Depending on the context, an object capable of bearing a truth value, an object of belief, or the meaning of a declarative sentence. In general, we use declarative sentences to express propositions, but not all declarative sentences actually do express propositions.

propositional logic The branch of logic that explores the logical relationships among propositions. Contrast with *Aristotelian logic* (or term logic).

quadriliteral diagram A sort of Venn diagram that can be used to represent the overlaps among four sets. See Section 5.1.

quality (of a proposition) Whether a categorical proposition is affirmative or negative.

quantity (of a proposition) Whether a categorical proposition is universal or particular.

realism In philosophy, realism with respect to some class of objects is the view that the objects actually exist and are not just useful fictions of some kind. In logic, it is generally the view that propositions, and therefore abstract objects, actually exist. Contrast with *nominalism*.

redundancy theory A view of truth in which asserting that a statement is true is precisely equivalent to asserting the statement itself, thereby making truth a redundant concept. Contrast with *coherence theory* and *correspondence theory*.

relevance logic Systems of logic that attempt to formalize a relevance relation between the parts of a conditional statement.

Retinends A term used by Lewis Carroll to refer to the terms from a set of categorical premises that appear in the conclusion drawn from those premises.

Russell's paradox The paradox that arises from defining the set S as "the set of all sets that are not elements of themselves." We get a contradiction from any attempt to determine whether or not S is an element of itself. See Section 16.2.

Sheffer stroke In classical logic, a binary connective that evaluates to true precisely when at least one of the two connected propositions is false. The Sheffer stroke corresponds to the English phrase "not both." It is of interest because all of the other connectives can be expressed using it alone. See Section 7.8.

sorites paradox A classic paradox that arises in situations where opposite ends of a continuum manifest some clear difference, but it is impossible to identify a clear point in the middle where the difference first occurs. See Section 13.4.

sorites puzzle A puzzle in which the solver is challenged to deduce the consequences of more than two categorical propositions. See Chapter 5.

Square of Opposition A traditional diagram, first considered by Aristotle, that shows the logical relationships among a statement and its converse, obverse, and contrapositive. See Figure 3.2.

subaltern A proposition Q that must be true when a given proposition P is true, and whose falsity implies the falsity of P. See Figure 3.2.

subcontraries Two propositions that cannot both be false but can both be true. See Figure 3.2.

Sudoku A popular type of logic puzzle. The solver must fill in each cell of a 9×9 grid with one of the digits from 1–9, in such a way that each row, column, and 3×3 block contains each digit exactly once. See Figure 2.1.

supposition, theory of In medieval logic, a classification of the various ways in which a written or spoken term might refer to some physical object. The term "supposition" is essentially synonymous with the modern term "reference."

syllogism An argument with two premises and a conclusion. See also *Aristotelian logic*.

symbolic language A purely formal language, usually invented for the purpose of illuminating logical relationships that might be vague and imprecise in natural languages. Contrast with *natural language*.

temporal logic A system of logic cognizant of notions of time.

term logic See *Aristotelian logic*.

triliteral diagram A sort of Venn diagram used by Lewis Carroll to represent the overlaps among three sets. See Section 4.2.

unexpected examination, the A classic epistemic paradox. A teacher tells his students they will have an examination the following week, but they will not know which day until they actually walk into class and receive the test. The students reason that the test cannot be on Friday, since if Thursday comes and goes with no test, they will know before class that the test is on Friday. Similar reasoning eliminates the other days of the week, and they assume the teacher was pulling their leg. Then they walk into class on Tuesday and receive the exam. Surprise! See Section 16.5.

universal affirmative/negative See *categorical proposition*.

universe of discourse The collection of all objects under consideration while undertaking some exercise of reasoning.

well-formed formula An expression in a formal language created in strict accordance with the syntactical rules of the system. Typically abbreviated "wff." The well-formed formulas of a formal language are analogous to the grammatically correct sentences of a natural language.

REFERENCES

Abeles, F. (2005). Lewis Carroll's Formal Logic. *History and Philosophy of Logic* **26**(1): 33–46.

Abeles, F. (2007). Lewis Carroll's Visual Logic. *History and Philosophy of Logic* **28**(1): 1–17.

Abeles, F. (2010). *The Pamphlets of Lewis Carroll*, vol. 4. Charlottesville: University Press of Virginia.

Abeles, F., and Moktefi, A. (2016). Correspondence with G. F. Stout, the Editor of *Mind. The Carrollian* (28): 9–13.

Alexander, P. (1969). *An Introduction to Logic: The Analysis of Arguments*. New York: Schocken Books.

Alexander, P. (1978). Review of *Symbolic Logic* by L. Carroll, edited by W. W. Bartley. *Philosophical Quarterly* **28**(113): 348.

Anonymous. (1887). Review of *The Game of Logic* by L. Carroll. *Literary Review* **16**(April): 122.

Aristotle. (2001). *The Basic Works of Aristotle*, R. McKean, ed. New York: Random House.

Arnauld, A., and Nicole, P. (1851). *The Port-Royal Logic*. T. S. Baynes, trans. Edinburgh: Sutherland and Knox.

Bach, K. (2002). Language, Logic, and Form. In: *A Companion to Philosophical Logic*, D. Jacquette, ed. Malden, MA: Wiley-Blackwell.

Bar-Hillel, Y. (1957). New Light on the Liar. *Analysis* **18**(1): 1–6.

Bartley, W. (1962). Achilles, the Tortoise, and Explanation in Science and History. *British Journal for the Philosophy of Science* **13**(49): 15–33.

Bartley, W. (1977). *Lewis Carroll's Symbolic Logic*. New York: Clarkson N. Potter.

Beall, J. C. (2008). *Revenge of the Liar: New Essays on the Paradox*. Oxford: Oxford University Press.

Beall, J. C., and Restall, G. (2005). *Logical Pluralism*. Oxford: Oxford University Press.

Beall, J. C., and Van Fraassen, B. C. (2003). *Possibilities and Paradox: An Introduction to Modal and Many-Valued Logic*. Oxford: Oxford University Press.

Bellos, A. (2017). *Can You Solve My Problems: Ingenious, Perplexing, and Totally Satisfying Math and Logic Puzzles*. New York: The Experiment.

Bennett, J. (2003). *A Philosophical Guide to Conditionals*. Oxford: Oxford University Press.

Besson, C. (2018). Norms, Reasons, and Reasoning: A Guide through Lewis Carroll's Regress Argument. In *The Oxford Handbook of Reasons and Normativity*, D. Star, ed. Oxford: Oxford University Press. 504–528.

Besson, C. (2020, anticipated). *The Limits of Cognitivism: Logic, Reasoning and the Tortoise*. Oxford: Oxford University Press.

Bewersdorff, J. (2004). *Luck, Logic, and White Lies: The Mathematics of Games*. Translated by D. Kramer. Wellesley, MA: A. K. Peters.

Bird, O. (1964). *Syllogistic and Its Extensions*. Englewood Cliffs, NJ: Prentice Hall.

Blackburn, S. (1995). Practical Tortoise Raising. *Mind* 104(416): 695–711.

Blanché, R. (1962). *Axiomatics*. London: Routledge and Kegan.

Boole, G. (1951). *An Investigation into the Laws of Thought, on Which Are Founded the Mathematical Theories of Logic and Probabilities*. Mineola, NY: Dover. Exact reproduction of the original 1854 edition.

Boolos, G. (1996). The Hardest Logic Puzzle Ever. *Harvard Review of Philosophy* 6(1): 62–65.

Bonevac, D. (2012). A History of Quantification. In *Handbook of the History of Logic*, Vol. 11, D. M. Gabbay, F. Pelletier, and J. Woods, eds. Amsterdam: North-Holland. 63–126.

Bonevac, D., and Dever, J. (2012). A History of the Connectives. In *Handbook of the History of Logic*, Vol. 11, D. M. Gabbay, F. Pelletier, and J. Woods, eds. Amsterdam: North-Holland. 175–233.

Burgess, J. (2009). *Philosophical Logic*. Princeton, NJ: Princeton University Press.

Burks, A. W., and Copi, I. M. (1950). Lewis Carroll's Barber Shop Paradox. *Mind* 59(234): 219–222.

Carnielli, W. (2017). Making the 'Hardest Logic Puzzle Ever' a Bit Harder. In *Raymond Smullyan on Self Reference*, M. Fitting and B. Rayman, eds. Cham, Switzerland: Springer. 181–190.

Carr, J. D. (1935). *The Three Coffins*. New York: Charter Books.

Carr, J. D. (1967). *Dark of the Moon*. New York: Berkley Medallion.

Carroll, L. (1894). A Logical Paradox. *Mind* 3(11): 436–438.

Carroll, L. (1895). What the Tortoise Said to Achilles. *Mind* 4(14): 278–280.

Carroll, L. (1958). *Symbolic Logic and the Game of Logic: Two Books Bound as One*. Mineola, NY: Dover. Reprint of the original 1896 and 1886 editions, respectively.

Cassiday, B. (ed.) (1983). *Roots of Detection: The Art of Deduction before Sherlock Holmes*. New York: Frederick Ungar.

Chandler, R. (1950). *The Simple Art of Murder*. New York: Houghton Mifflin.

Chihara, C. S. (1985). Olin, Quine, and the Surprise Examination. *Philosophical Studies* 47(2): 191–199.

Chow, T. Y. (1998). The Surprise Examination or Unexpected Hanging Paradox. *American Mathematical Monthly* 105(1): 41–51.

Christie, A. (1983). *Murder on the Links*. New York: Dell.

Clark, M. (2012). *Paradoxes from A to Z*, 3rd ed. New York: Routledge.

Cook, R. T. (2010). Let a Thousand Flowers Bloom: A Tour of Logical Pluralism. *Philosophy Compass* 5(6): 492–504.

Delahaye, J.-P. (2006). The Science behind Sudoku. *Scientific American* (June): 80–87.

Doyle, A. C. (1975a). *The Memoirs of Sherlock Holmes*. New York: Ballantine Books.

Doyle, A. C. (1975b). *A Study in Scarlet*. New York: Ballantine Books.

Edis, T. (1998). How Gödel's Theorem Supports the Possibility of Machine Intelligence. *Minds and Machines* 8(2): 251–262.

Eklund, M. (2002). Deep Inconsistency. *Australasian Journal of Philosophy* 80(3): 321–331.

Eklund, M. (2012). The Multitude View on Logic. In *New Waves in Philosophical Logic*, G. Restall and G. Russell, eds. New York: Palgrave Macmillan.

Ellison, H. (2003). Introduction to *Jacques Futrelle's Thinking Machine*. New York: Modern Library.

Engel, P. (2016). The Philosophical Significance of Carroll's Regress. *The Carrollian* (28): 84–111.

Field, H. (2009). Pluralism in Logic. *Review of Symbolic Logic* 2(2): 342–359.

Fitting, M., and Rayman, B. (2017). *Raymond Smullyan on Self-Reference*. Cham, Switzerland: Springer.

Flage, D. (1994). *Understanding Logic*. Englewood Cliffs, NJ: Prentice Hall.

Floridi, L. (1997). Skepticism and Animal Rationality: The Fortune of Chrysippus's Dog in the History of Western Thought. *Archiv für Geschichte der Philosophie* 79: 27–57.

Ford, L. R. (1947). Problem E776. *American Mathematical Monthly* 54(6): 339–340.

Frank, F. S., and Magistrale, A. (1997). *The Poe Encyclopedia*. Westport, CT: Greenwood.

Franzén, T. (2005). *Gödel's Theorem: An Incomplete Guide to Its Use and Abuse*. Wellesley, MA: A. K. Peters.

Freeman, R. A. (2010). *The Red Thumb Mark*. San Luis Obispo, CA: Resurrected Press.

Futrelle, J. (2003). *Jacques Futrelle's The Thinking Machine: The Enigmatic Problems of Prof. Augustus S. F. Van Dusen*. New York: Random House.

Gardner, M. (1991). *The Unexpected Hanging and Other Mathematical Diversions*. Chicago: University of Chicago Press.

Gardner, M. (2006). *The Colossal Book of Short Puzzles and Problems*. New York: W. W. Norton.

Garver, N. (1970). The Range of Truth and Falsity. In *The Paradox of the Liar*, R. L. Martin, ed. New Haven, CT: Yale University Press. 121–126.

George, R., and Van Evra, J. (2002). The Rise of Modern Logic. In *A Companion to Philosophical Logic*, D. Jacquette, ed. Malden, MA: Blackwell.

Goldstein, R. (2006). *Incompleteness: The Proof and Paradox of Kurt Gödel*. New York: W. W. Norton.

Goodman, N. (1972). *Problems and Projects*. Indianapolis, IN: Bobbs Merrill.

Gottwald, S. (2015). Many-Valued Logics. In: *The Stanford Encyclopedia of Philosophy*. https://plato.stanford.edu/entries/logic-manyvalued/#AppHarDes. Last accessed July 2020.

Gowers, T. (2012). Vividness in Mathematics and Narrative. In: *Circles Disturbed: The Interplay of Mathematics and Narrative*, A. Doxiadis and B. Mazur, eds. Princeton, NJ: Princeton University Press. 211—232.

Grayling, A. C. (1982). *An Introduction to Philosophical Logic*. Sussex, UK: Harvester Press.

Haack, S. (1974). *Deviant Logic: Some Philosophical Issues*. Cambridge: Cambridge University Press.

Haack, S. (1978). *Philosophy of Logics*. Cambridge: Cambridge University Press.

Haining, P., ed. (1996). *Hunted Down: The Detective Stories of Charles Dickens*. London: Peter Owen.

Harman, G. (1984). Logic and Reasoning. *Synthese* 60(1): 107–127.

Imholtz, C., and Moktefi, A. (2016). What the Tortoise Said to Achilles: A Selective Bibliography. *The Carrollian* (28): 128–136.

Johnson, W. E. (1894). A Logical Paradox. *Mind* 3(12): 583.

Jones, E.E.C. (1905). Lewis Carroll's Logical Paradox. *Mind* 14(53): 146–148.

Kant, I. (2003). *Critique of Pure Reason*, 2nd ed. N. K. Smith, trans. New York: Palgrave Macmillan.

Kemelman, H. (1967). *The Nine Mile Walk*. Greenwich: Fawcett Crest.

Kilwardby, R. (1988). *De Ortu Scientiarum*. In *The Cambridge Translations of Medieval Philosophical Texts*, Vol. 1, N. Kreitzmann and E. Stump, eds. Cambridge: Cambridge University Press.

Kissel, T. K. (2016). Logical Instrumentalism. Ph.D. thesis, Ohio State University, Columbus.

Kneale, W. (1971). Russell's Paradox and Some Others. *British Journal for the Philosophy of Science* 22(4): 321–338.

Kneale, W. (1972). Propositions and Truth in Natural Languages. *Mind* 81(322): 225–243.

Kneale, W., and Kneale, M. (1962). *The Development of Logic*. London: Oxford University Press.

Lacey, A. R. (1970). The Paradox of the Preface. *Mind* 79(316): 614–615.

Lear, J. (1980). *Aristotle and Logical Theory*. Cambridge: Cambridge University Press.

Leszl, W. (2004). Aristotle's Logical Works and His Conception of Logic. *Topoi* 23(1): 71–100.

Lewis, C. I. (1932). Alternative Systems of Logic. *The Monist* 42(4): 481–507.

Lindberg, D. (2009). Myth 1: That the Rise of Christianity was Responsible for the Demise of Ancient Science. In *Galileo Goes to Jail and Other Myths about Science and Religion*, R. Numbers, ed. Cambridge, MA: Harvard University Press.

Locke, J. (1979). *An Essay Concerning Human Understanding*. Oxford: Oxford University Press.

Makinson, D. C. (1965). The Paradox of the Preface. *Analysis* 25(6): 205–207.

Martin, R. L. (1970). *The Paradox of the Liar*. New Haven, CT: Yale University Press.

Mill, J. S. (1974). *A System of Logic, Ratiocinative and Inductive: Being a Connected View of the Principles of Evidence and the Methods of Investigation*. Vol. VII of *The Collected Works of John Stuart Mill*, J. M. Robson, gen. ed. Toronto: University of Toronto Press.

Milton, J. (1644). Of Education. Online at www.dartmouth.edu/~milton/reading_room/of_education/text.shtml. Last accessed July 2020.

Moktefi, A. (2008). Lewis Carroll's Logic. In *British Logic in the Nineteenth Century*, D. M. Gabbay and J. Woods, eds. Amsterdam: North-Holland.

Moktefi, A. (2013). Beyond Syllogisms: Carroll's (Marked) Quadriliteral Diagram. In *Visual Reasoning with Diagrams*, A. Moktefi and S.-J. Shin, eds. Basel: Birkhäuser.

Nagel, E., and Newman, J. R. (2001). *Gödel's Proof*, rev. ed. New York: New York University Press.

New, C. (1978). A Note on the Paradox of the Preface. *Philosophical Quarterly* 28(113): 341–344.

Okasha, S. (2002). *Philosophy of Science: A Very Short Introduction*. Oxford: Oxford University Press.

O'Toole, R. R., and Jennings, R. E. (2004). The Megarians and the Stoics. In *The Handbook for the History of Logic*, Vol. 1, D. M. Gabbay and J. Woods, eds. Amsterdam: North-Holland, 397–522.

Peckhaus, V. (2018). Leibniz's Influence on 19th Century Logic. In *The Stanford Encyclopedia of Philosophy*. https://plato.stanford.edu/entries/leibniz-logic-influence/. Last accessed July 2020.

Pigliucci, M. (2012). Sherlock's Reasoning Toolbox. In *The Philosophy of Sherlock Holmes*, P. Tallon and D. Baggett, eds. Lexington: University Press of Kentucky.

Poe, E. A. (2006). *The Murders in the Rue Morgue: The Dupin Tales*. New York: Modern Library.

Priest, G. (2006). *In Contradiction*, 2nd. ed. Oxford: Oxford University Press.

Priest, G. (2008a). *Doubt Truth to Be a Liar*. Oxford: Oxford University Press.

Priest, G. (2008b). *An Introduction to Non-Classical Logic: From If to Is*. Cambridge: Cambridge University Press.

Priest, G., Beall, J. C., and Armour-Garb, B. (2004). *The Law of Non-Contradiction: New Philosophical Essays*. Oxford: Oxford University Press.

Prior, A. N. (1958). Epimenides and the Cretan. *Journal of Symbolic Logic* 23(3): 261–266.

Putnam, H. (1971). *Philosophy of Logic*. Oxon, UK: George Allen and Unwin.

Queen, E. (1930). *The French Powder Mystery*. New York: International Readers League.

Quine, W.V.O. (1953). On a So-Called Paradox. *Mind* 62(245): 65–67.

Quine, W.V.O. (1958). *Mathematical Logic*. Cambridge, MA: Harvard University Press.

Rabern, B., and Rabern, L. (2008). A Simple Solution to the Hardest Logic Puzzle Ever. *Analysis* 68(2): 105–112.

Rasmussen, J. (2014). *Defending the Correspondence Theory of Truth*. Cambridge: Cambridge University Press.

Read, S. (1995). *Thinking about Logic: An Introduction to the Philosophy of Logic*. Oxford: Oxford University Press.

Rescher, N. (1969). *Many-Valued Logic*. New York: McGraw-Hill.

Roberts, T. S. (2001). Some Thoughts about the Hardest Logic Puzzle Ever. *Journal of Philosophical Logic* 30(6): 609–612.

Rose, L. (1968). *Aristotle's Syllogistic*. Springfield, IL: Charles C. Thomas.

Rosenhouse, J. (2009). *The Monty Hall Problem: The Math behind the World's Most Contentious Brainteaser*. New York: Oxford University Press.

Rosenhouse, J. (2014). *Four Lives: A Celebration of Raymond Smullyan*. Mineola: Dover.

Rosenhouse, J. (2014). Knights, Knaves, Normals, and Neutrals. *College Mathematics Journal* 45(4): 297–306.

Rosenhouse, J. (2016). Fuzzy Knights and Knaves. *Mathematics Magazine* 89(4): 268–280.

Rosenhouse, J., and Taalman, L. (2012). *Taking Sudoku Seriously: The Math behind the World's Most Popular Pencil Puzzle*. New York: Oxford University Press.

Russell, B. (1902). The Teaching of Euclid. *Mathematics Gazette* 2(33): 165–167.

Russell, B. (1903). *The Principles of Mathematics*. New York: W. W. Norton.

Russell, B. (1908). Mathematical Logic as Based on the Theory of Types. *American Journal of Mathematics* 30(3): 222–262.

Russell, B. (1972). *The History of Western Philosophy*. New York: Simon and Schuster. (Reprint of the original 1945 edition).

Ryle, G. (1945–1946). Knowing How and Knowing That. *Proceedings of the Aristotelian Society* 46: 1–16.

Ryle, G. (1950). If, So, and Because. In *Philosophical Analysis*, M. Black, ed. Englewood Cliffs, NJ: Prentice Hall.

Sainsbury, R. M. (2009). *Paradoxes*, 3rd ed. Cambridge: Cambridge University Press.

Shapiro, S. (2004). Simple Truth, Contradiction, and Consistency. In *The Law of Non-Contradiction: New Philosophical Essays*. G. Priest, J. C. Beall, and B. Armour-Garb, eds. Oxford: Oxford University Press.

Shenefelt, M., and White, H. (2013). *If A Then B: How the World Discovered Logic*. New York: Columbia University Press.

Sheffer, H. M. (1913). A Set Five Independent Postulates for Boolean Algebras, with Application to Logical Constants. *Transactions of the American Mathematical Society*. 14(4): 481–88.

Sidgwick, A. (1894). A Logical Paradox. *Mind* 3(12): 582.

Smith, N.J.J. (2012). *Logic: The Laws of Truth*. Princeton, NJ: Princeton University Press.

Smith, R. (2017). Aristotle's Logic. In: *The Stanford Encyclopedia of Philosophy*, E. N. Zalta, ed. https://plato.stanford.edu/entries/aristotle-logic/. Last accessed July 2020.

Smullyan, R. (1978). *What Is the Name of This Book? The Riddle of Dracula and Other Logical Puzzles*. Englewood Cliffs, NJ: Prentice Hall.

Smullyan, R. (1979). *The Chess Mysteries of Sherlock Holmes*. New York: Knopf.

Smullyan, R. (1980). *This Book Needs No Title: A Budget of Living Paradoxes*. New York: Simon and Schuster.

Smullyan, R. (1981). *The Chess Mysteries of the Arabian Knights*. New York: Knopf.

Smullyan, R. (1982a). *Alice in Puzzle-Land: A Carrollian Tale for Children under Eighty*. Mineola, NY: Dover.

Smullyan, R. (1982b). *The Lady or the Tiger, and Other Logic Puzzles*. New York: Knopf.

Smullyan, R. (1983). *5000 B.C. and Other Philosophical Fantasies: Puzzles and Paradoxes, Riddles and Reasonings*. New York: St. Martin's Press.

Smullyan, R. (1985). *To Mock a Mockingbird, and Other Logic Puzzles*. New York: Knopf.

Smullyan, R. (1987). *Forever Undecided: A Puzzle Guide to Gödel*. New York: Knopf.

Smullyan, R. (1992a). *Satan, Cantor, and Infinity and Other Mind-Boggling Puzzles*. New York: Knopf.

Smullyan, R. (1992b). *Gödel's Incompleteness Theorems*. New York: Oxford University Press.

Smullyan, R. (1994). *Diagonalization and Self-Reference*. Oxford: Oxford University Press.

Smullyan, R. (1995). *First Order Logic*. Reprint of the original 1968 edition, published by Springer. Mineola, NY: Dover.

Smullyan, R. (1997). *The Riddle of Scheherazade and Other Amazing Puzzles*. New York: Knopf.

Smullyan, R. (2009). *Logical Labyrinths*. Wellesley, MA: A. K. Peters.

Smullyan, R. (2010). *King Arthur in Search of His Dog*. Mineola, NY: Dover.

Smullyan, R. (2013). *The Gödelian Puzzle Book: Puzzles, Paradoxes, and Proofs*. Mineola, NY: Dover.

Smullyan, R. (2015). *Reflections: The Magic, Music and Mathematics of Raymond Smullyan*. Singapore: World Scientific.

Sorensen, R. (2003). *A Brief History of the Paradox: Philosophy and the Labyrinths of the Mind*. New York: Oxford University Press.

Steinberger, F. (2016). The Normative Status of Logic. In *The Stanford Encyclopedia of Philosophy*, E.N. Zalta, ed. https://plato.stanford.edu/entries/logic-normative/. Last accessed July 2020.

Stroud, B. (1979). Inference, Belief, and Understanding. *Mind* 88(350): 179–196.

Thomson, J. F. (1960). What Achilles Should Have Said to the Tortoise. *Ratio* 3: 95–105.

Toulmin, S. (1953). *The Philosophy of Science: An Introduction*. London: Hutchinson's University Library.

Uzquiano, G. (2010). How to Solve the Hardest Logic Puzzle Ever in Two Questions. *Analysis* 70(1): 39–44.

Van Ditmarsch, H., and Kooi, B. (2015). *100 Prisoners and a Light Bulb*. New York: Springer.

Venn, J. (1881). *Symbolic Logic*. London: Macmillan and Co.

Voltaire. (1966). *Candide and Other Stories*. London: Oxford University Press.

Wheeler, G., and Barahona, P. (2012). Why the Hardest Logic Puzzle Ever Cannot Be Solved in Less Than Three Questions. *Journal of Philosophical Logic* 41(2): 493–503.

Wieland, J. W. (2013). What Carroll's Tortoise Actually Proves. *Ethical Theory and Moral Practice* 16(5): 983–997.

Wilks, I. (2008). Peter Abelard and His Contemporaries. In *The Handbook for the History of Logic*, Vol. 2, D. M. Gabbay and J. Woods, eds. Amsterdam: North-Holland.

Williams, J. N. (1987). The Preface Paradox Dissolved. *Theoria* 53(2–3): 121–140.

Wilson, F. (2008). The Logic of John Stuart Mill. In *The Handbook for the History of Logic*, Vol. 4, D. M. Gabbay and J. Woods, eds. Amsterdam: North-Holland.

Wilson, J. C. (1905). Lewis Carroll's Logical Paradox. *Mind* 14(54): 292–293.

Wilson, R., and Moktefi, A. (2019). *The Mathematical World of Charles Dodgson*. Oxford: Oxford University Press.

Wintein, S. (2012). On the Behavior of True and False. *Minds and Machines* 22(1): 1–24.

Wright, C., and Sudbury, A. (1977). The Paradox of the Unexpected Examination. *Australasian Journal of Philosophy* 55(1): 41–58.

INDEX